# Lecture Notes in Computer Science 12559

More information about this subseries at http://www.springer.com/series/7407

Nenad Mladenovic · Andrei Sleptchenko ·
Angelo Sifaleras · Mohammed Omar (Eds.)

# Variable
# Neighborhood Search

8th International Conference, ICVNS 2021
Abu Dhabi, United Arab Emirates, March 21–25, 2021
Proceedings

 Springer

*Editors*
Nenad Mladenovic (iD)
Khalifa University
Abu Dhabi, United Arab Emirates

Andrei Sleptchenko (iD)
Khalifa University
Abu Dhabi, United Arab Emirates

Angelo Sifaleras (iD)
University of Macedonia
Thessaloniki, Greece

Mohammed Omar (iD)
Khalifa University
Abu Dhabi, United Arab Emirates

ISSN 0302-9743                    ISSN 1611-3349    (electronic)
Lecture Notes in Computer Science
ISBN 978-3-030-69624-5           ISBN 978-3-030-69625-2    (eBook)
https://doi.org/10.1007/978-3-030-69625-2

LNCS Sublibrary: SL1 – Theoretical Computer Science and General Issues

This Springer imprint is published by the registered company Springer Nature Switzerland AG
The registered company address is: Gewerbestrasse 11, 6330 Cham, Switzerland

# Preface

This volume edited by Nenad Mladenović, Andrei Sleptchenko, Angelo Sifaleras, and Mohammed Omar contains peer-reviewed papers from the 8th International Conference on Variable Neighborhood Search (ICVNS 2021) held in Abu Dhabi, U.A.E., during March 21–25, 2021.

The conference follows previous successful meetings that were held in Puerto de la Cruz, Tenerife, Spain (2005); Herceg Novi, Montenegro (2012); Djerba, Tunisia (2014); Malaga, Spain (2016); Ouro Preto, Brazil, (2017); Sithonia, Halkidiki, Greece (2018); Rabat, Marocco (2019).

This edition was organized by Nenad Mladenović, Andrei Sleptchenko, and Mohammed Omar from Khalifa University (United Arab Emirates), together with Angelo Sifaleras, from the University of Macedonia (Greece).

Like its predecessors, the main goal of ICVNS 2021 was to provide a stimulating environment in which researchers coming from various scientific fields could share and discuss their knowledge, expertise, and ideas related to the VNS Metaheuristic and its applications. Due to the COVID-19 pandemic, the ICVNS 2021 was organized in hybrid (online and offline) mode with the help of the Office of Marketing and Communications of Khalifa University.

The following three plenary lecturers shared their current research directions with the ICVNS 2021 participants:

- Panos M. Pardalos, from the Center for Applied Optimization, Department of Industrial and Systems Engineering, of the University of Florida, USA,
- Yury Kochetov, from the Sobolev Institute of Mathematics, Novosibirsk, Russia,
- Bassem Jarboui, from the Higher Colleges of Technology, U.A.E.

Around 40 participants took part in the ICVNS 2021 conference, and a total of 27 papers were accepted for oral presentation. A total of 14 long papers were accepted for publication in this LNCS volume after thorough peer review by the members of the ICVNS 2021 Program Committee. These papers describe recent advances in methods and applications of Variable Neighborhood Search.

The editors thank all the participants in the conference for their contributions and for their continuous effort to disseminate VNS and are grateful to the reviewers for preparing excellent reports. The editors wish to acknowledge the Springer LNCS editorial staff for their support during the entire process of making this volume. Finally, we express our gratitude to the organizers and sponsors of the ICVNS 2021 meeting:

- The Research Center for Digital Supply Chains and Operations Management, Khalifa University,
- the Office of Marketing and Communications of Khalifa University

- The EURO Working Group on Metaheuristics (EWG EU/ME),
- The Department of Culture and Tourism of Abu Dhabi.

Their support is greatly appreciated for making ICVNS 2021 a great scientific event.

January 2021

Nenad Mladenovic
Andrei Sleptchenko
Angelo Sifaleras
Mohammed Omar

# Organization

## Program Chairs

| | |
|---|---|
| Nenad Mladenovic | Khalifa University, United Arab Emirates |
| Andrei Sleptchenko | Khalifa University, United Arab Emirates |
| Angelo Sifaleras | University of Macedonia, Greece |
| Mohammed Omar | Khalifa University, United Arab Emirates |

## Program Committee

| | |
|---|---|
| Ali Allahverdi | Kuwait University, Kuwait |
| Daniel Aloise | GERAD, Polytechnique Montréal, Canada |
| Ada Alvarez | Universidad Autónoma de Nuevo León, Mexico |
| Cláudio Alves | Universidade do Minho, Portugal |
| John Beasley | Brunel University London, UK |
| Abdelghani Bekrar | Université Polytechnique Hauts-de-France, France |
| Rachid Benmansour | INSEA, Morocco |
| Jack Brimberg | Royal Military College of Canada, Canada |
| Mirjana Cangalović | University of Belgrade, Serbia |
| Gilles Caporossi | HEC Montréal, Canada |
| Emilio Carrizosa | Universidad de Sevilla, Spain |
| Silvia Casado | Universidad de Burgos, Spain |
| Sergio Consoli | European Commission, Italy |
| Teodor G. Crainic | Université du Québec à Montréal, Canada |
| Abraham Duarte | Universidad Rey Juan Carlos, Spain |
| Karl Dörner | Universität Wien, Austria |
| Anton Eremeev | Omsk State University, Russia |
| Laureano Escudero | Universidad Rey Juan Carlos, Spain |
| Adriana Gabor | Khalifa University, United Arab Emirates |
| Haroldo Gambini Santos | Universidade Federal de Ouro Preto, Brazil |
| Michel Gendreau | École Polytechnique de Montréal, Canada |
| Saïd Hanafi | Université Polytechnique Hauts-de-France, France |
| Pierre Hansen | HEC Montréal, Canada |
| Chandra Irawan | Nottingham University Business School China, China |
| Bassem Jarboui | Higher Colleges of Technology, United Arab Emirates |
| Yury Kochetov | Novosibirsk State University, Russia |
| Vera Kovacevic Vujcic | University of Belgrade, Serbia |
| Leo Liberti | CNRS & École Polytechnique, France |
| Jianyi Lin | Khalifa University, United Arab Emirates |
| Yannis Marinakis | Technical University of Crete, Greece |
| Belén Melián | Universidad de La Laguna, Spain |
| Athanasios Migdalas | Luleå University of Technology, Sweden |

| | |
|---|---|
| Nenad Mladenovic | Khalifa University, United Arab Emirates |
| José Andrés Moreno Pérez | Universidad de La Laguna, Spain |
| Vitor Nazário Coelho | OptBlocks, Brazil |
| Luiz Satoru Ochi | Fluminense Federal University (UFF), Brazil |
| Mohammed Omar | Khalifa University, United Arab Emirates |
| Joaquín Pacheco | University of Burgos, Spain |
| Panos Pardalos | University of Florida, USA |
| Eduardo G. Pardo | Universidad Rey Juan Carlos, Spain |
| Jun Pei | Hefei University of Technology, China |
| Leonidas Pitsoulis | Aristotle University of Thessaloniki, Greece |
| Günther Raidl | TU Wien, Austria |
| Celso Ribeiro | Universidade Federal Fluminense, Brazil |
| Said Salhi | University of Kent, UK |
| Nikolaos Samaras | University of Macedonia, Greece |
| Patrick Siarry | Université Paris-Est Créteil, France |
| Angelo Sifaleras | University of Macedonia, Greece |
| Andrei Sleptchenko | Khalifa University, United Arab Emirates |
| Marcone Souza | Universidade Federal de Ouro Preto, Brazil |
| Thomas Stützle | Université Libre de Bruxelles (ULB), Belgium |
| Kenneth Sörensen | University of Antwerp, Belgium |
| Christos Tarantilis | Athens University of Economics and Business, Greece |
| Hasan Turan | University of New South Wales, Australia |
| Dragan Urošević | Mathematical Institute SANU, Serbia |

# Contents

# Finding Critical Nodes in Networks Using Variable Neighborhood Search

Iván Martín de San Lázaro[iD], Jesús Sánchez-Oro[✉][iD], and Abraham Duarte[iD]

Universidad Rey Juan Carlos, C/Tulipán S/N, Móstoles, Spain
{ivan.martin,jesus.sanchezoro,abraham.duarte}@urjc.es

**Abstract.** Several problems related to networks are based on the identification of certain nodes which can be relevant for different tasks: network security and stability, protein interaction, or social influence analysis, among others. These problems can be modeled with the Critical Node Detection Problem (CNDP). Given a network, the CNDP consists of identifying a set of $p$ nodes whose removal minimizes the pairwise connectivity of the network. In this work, a Basic Variable Neighborhood Search (BVNS) algorithm is presented with the aim of generating high quality solutions in short computing times. The detailed experimental results show the performance of the proposed algorithm when comparing it with the state of the art method, emerging BVNS as a competitive algorithm for the CNDP.

**Keywords:** Critical Node Detection Problem · Variable Neighborhood Search · Constructive procedure · Metaheuristics

## 1  Introduction

The stability of a network usually relies on a small set of the nodes that conforms it, which can be labeled as critical nodes. A node can be considered critical if the network performance is deteriorated when the node fails or it is affected by an external attack. Therefore, identifying these critical nodes is a relevant task which has been the focus of researchers and practitioners in the last years. These critical nodes can be found in the literature under different names: most vital nodes [11], key-player nodes [8], most influential nodes [19], or most k-mediator nodes [21], among others. In the context of network connectivity, in which this paper is focused, they are usually known as critical nodes [20].

Identifying critical nodes in networks can be useful in several tasks. In computational biology, several biological organisms are interconnected proteins that interact among them to form a protein-interaction network. These networks are usually modeled by a graph where nodes are proteins and edges are interactions between them. Critical nodes are proteins that maintain connectivity among all proteins, and they can provide useful information for many biological applications. For instance, in drug design [18, 22], these critical proteins are the objective to neutralize the studied harmful organism [7, 30].

© Springer Nature Switzerland AG 2021
N. Mladenovic et al. (Eds.): ICVNS 2021, LNCS 12559, pp. 1–13, 2021.
https://doi.org/10.1007/978-3-030-69625-2_1

In the context of network security, the CNDP is able to find relevant vulnerabilities in a network. In general, a vulnerability in a network usually consists in a node failure that deteriorates the network performance. Therefore, identifying critical nodes in a network can allow us to reinforce them in order to guarantee the network performance [8,14]. For instance, a method based on the critical nodes concept [15] is presented for analyzing network vulnerability in the case of unexpected events.

Identifying critical nodes is also interesting for analyzing telecommunication networks, since the failure of these nodes can lead to disable the network [25,26]. From the attacker point of view, the CNP has been studied to destroy communications on terrorist networks [5]. The Wireless Network Jamming Problem was also formulated as a CNDP variant [10], where critical nodes are the ones that need to be jammed.

The CNDP is also interesting in transportation problems, where the managers need to analyze the relevance of certain points in the road network [9]. Additionally, identifying critical nodes can help to prevent from disasters by planning good emergency evacuations [34].

The problem of identifying these critical nodes in networks is usually known as Critical Node Detection Problem (CNDP). The CNDP is an optimization problem that aims to find the set of nodes in a network that leads to maximize or minimize a certain criteria related to the network connectivity. Since there exists different objective function to be optimized in the context of critical node detection, we will firstly describe the objective function considered in this work.

Let $G = (V, E)$ be a network modeled as an undirected and unweighted graph where $V$ is the set of nodes, with $|V| = n$, and $E$ is the set of edges, with $|E| = m$. The objective of the CNDP considered in this work is to find a subset of nodes $S$, with $S \subseteq S$, of size $p$, whose removal leads to a graph $G' = (V \setminus S, E')$, with $E' = E \setminus \{(u, v) \in E : u \in S\}$ with the minimum pairwise connectivity. Then, the graph $G'$ can be represented by a set of connected components $G' = \{C_1, C_2, \ldots, C_c\}$. Notice that each pair of nodes in a connected component $C_i$ is connected through a path, then existing $\frac{|C_i| \cdot (|C_i|-1)}{2}$ paths in it. A solution $S$ for the CNDP is then evaluated as the number of paths between two nodes that exists in the graph $G'$ that results after removing all nodes in $S$ from the original graph $G$. More formally,

$$CNDP(S) = \sum_{i=1}^{c} \frac{|C_i| \cdot (|C_i| - 1)}{2}$$

Then, the CNDP consists of finding a solution $S^\star$ with the minimum $CNDP(S)$ value among all possible sets of $p$ nodes, $S$, that can be conformed from $V$. In mathematical terms,

$$S^\star = \arg\min_{S \in \mathcal{S}} CNDP(S)$$

Notice that, given a network with $n$ nodes and a certain value of $p$, the number of solutions in the search space can be evaluated as $\binom{n}{p}$. Since this value increases extremely fast with the $n$ and $p$ values, and the real-life networks are usually large, it is necessary to propose new efficient algorithms that provides high quality solutions in small computing times.

Figure 1(a) shows an example network $G$ with 12 nodes and 16 edges, as well as two feasible solutions $S_1$ and $S_2$ for the CNDP considering $p = 2$. The initial network is represented in Fig. 1(a). In both presented solutions the selected nodes are highlighted in black and the edges that are removed as a consequence of this critical node selection are represented with a dashed line. Additionally, we have highlighted in gray the resulting connected components after removing the critical nodes.

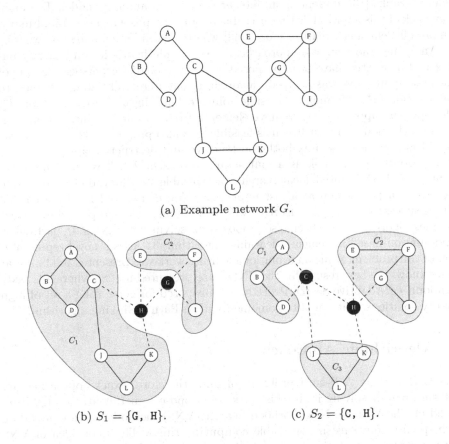

(a) Example network $G$.

(b) $S_1 = \{\texttt{G}, \texttt{H}\}$.    (c) $S_2 = \{\texttt{C}, \texttt{H}\}$.

**Fig. 1.** Example network with 12 nodes and 16 edges and two feasible and two feasible solutions for the CNDP, $S_1$ and $S_2$.

Solution depicted in Figure 1(b) is $S_1 = \{\texttt{G}, \texttt{H}\}$, resulting in two connected components, $C_1$ and $C_2$, with 7 and 3 nodes, respectively. Therefore, $CNDP(S_1) = \frac{7 \cdot 6}{2} + \frac{3 \cdot 2}{2} = 24$, i.e., there exists a path between 24 pairs of nodes in the resulting graph. If we now evaluate solution $S_2$, presented in Fig. 1(c), there are three connected components, $C_1$, $C_2$, and $C_3$, with sizes 3, 4, and 3, respectively. The objective function value for $S_2$ is $CNDP(S_2) = \frac{3 \cdot 2}{2} + \frac{4 \cdot 3}{2} + \frac{3 \cdot 2}{2} = 12$. Therefore, analyzing these results, $S_2$ is better than $S_1$, since it is able to find a solution that allows a smaller number of pairwise connections than $S_2$.

The CNDP has been widely studied in the literature from both exact and heuristic perspectives. Due to the problem complexity, the exact approaches are designed for finding the optimal value in networks with a particular topology. For instance, the CNDP was proven to be polynomially solvable for trees using dynamic programming [12], and this result was later generalized for graphs with a bounded treewidth [2]. A branch and cut algorithm was also proposed for general graphs [13], presenting an integer linear programming model. The main drawback of this approach is that it is able to solve graphs when $n \leq 150$. Finally, a more efficient mathematical model [33] was tested in larger sparse networks.

Analyzing the heuristic approaches, both simple heuristics and more complex metaheuristics have been proposed. From the heuristic perspective, one of the first approaches was a greedy algorithm that iteratively adds elements to the solution until obtaining a feasible solution following a greedy criterion [6]. The opposite approach, i.e., remove elements from an initial solution that contains all the nodes until it becomes feasible was also proposed [32]. Later, some hybrid approaches that mix both constructive and destructive approximations were proposed [1], as well as a multi-start approach [27]. Two local improvements based on iterated local search and variable neighborhood search were presented in [4], as well as a fast implementation of iterated local search [35] whose success relies in an effective two-phase node exchange procedure. More complex algorithms have been also proposed for solving the CNDP. In particular, a population-based incremental learning algorithm [31] uses a novel representation of the problem, while an evolutionary framework was presented in [3]. As far as we know, the best results for the CNDP are presented in [28], where a Greedy Randomized Adaptive Search Procedure is proposed for generating a set of high quality solutions that are later combined with a Path Relinking algorithm.

## 2    Algorithmic Approach

The CNDP is an $\mathcal{NP}$-hard problem [6] and, therefore, exact approaches are not suitable for solving it. In this work we propose a metaheuristic algorithm based on the Variable Neighborhood Search (VNS) methodology for providing high quality solutions in reasonable computing times. The main idea of VNS is to perform systematic changes in the neighborhood structures in order to avoid stagnating the search in attraction basins. VNS was initially proposed as a simple metaheuristic algorithm [23], but it has evolved until becoming a complete framework with several extensions [17]: Basic VNS, Reduced VNS, General VNS, Variable Neighborhood Descent, and Variable Formulation Search, among others. This work is focused in the Basic VNS (BVNS) variant, since the main objective is to generate competitive solutions without requiring high computational efforts. BVNS performs both random (diversification) and deterministic (intensification) changes of neighborhood structures with the aim of avoiding getting stuck in local optima during the search. Algorithm 1 presents the pseudocode of the proposed BVNS.

---

**Algorithm 1.** $BVNS(S, k_{\max}, T)$

---

1: **for** $i \in 1 \ldots T$ **do**
2:     $k \leftarrow 1$
3:     **while** $k \leq k_{\max}$ **do**
4:         $S' \leftarrow Shake(S, k)$
5:         $S'' \leftarrow Improve(S')$
6:         $k \leftarrow NeighborhoodChange(S, S'', k)$
7:     **end while**
8: **end for**
9: **return** $S$

---

BVNS requires three input parameters: an initial feasible solution $S$ to start the search, the maximum neighborhood to be explored $k_{\max}$, and the maximum number of iterations of BVNS that will be performed, $T$. In the context of VNS, the initial solution can be generated either at random or using a more elaborated constructive procedure. Starting the search from a promising solution usually allows the algorithm to reduce the computational effort required to converge and, therefore, we propose a constructive procedure for the CNDP (see Sect. 2.1). We refer the reader to Sect. 3 where the impact of $k_{\max}$ and $T$ is thoroughly discussed.

BVNS starts from the first neighborhood (step 2), iterating until reaching the maximum predefined neighborhood (steps 3–7). In each iteration, the incumbent solution $S$ is randomly perturbed to find a neighbor solution $S'$ in the neighborhood $k$ under exploration (step 4), by using a *Shake* procedure which is described in Sect. 2.2. Since $S'$ has been selected at random from the current neighborhood, it is not necessarily a local optimum with respect to the neighborhood considered in the improvement procedure. Therefore, a local improvement is applied to $S'$, producing local optimum $S''$ (step 5). The local improvement proposed in this work is deeply described in Sect. 2.3. Finally, the neighborhood change method (see Sect. 2.4) is responsible for selecting the next neighborhood to be explored, as well as updating the incumbent solution if necessary (step 6). A complete iteration ends when reaching the largest neighborhood to be explored, $k_{\max}$. The BVNS ends when performing $T$ complete iterations, returning the best solution found during the search, $S$ (step 9).

## 2.1 Initial Solution

This work proposes a fast greedy constructive procedure to generate an promising solution that will be the starting point of the BVNS. The proposed greedy procedure starts from an empty solution $S$ and iteratively adds to $S$ the most promising node in each iteration. In order to select which is the most promising node, a greedy function is necessary, which evaluates each candidate node before inserting it in the solution.

Analyzing the CNDP, a node is a good candidate if it is able to disconnect several paths. Therefore, we propose to use an adaptation of a centrality measure derived from Social Network Analysis, called betweenness [16]. Given a vertex $v$, the betweenness centrality $b(v)$ is evaluated as

$$b(v) = \sum_{s,t \in V \setminus \{v\}} \frac{\sigma(s,t|v)}{\sigma(s,t)}$$

where $\sigma(s,t|v)$ is the number of paths between $s$ and $t$ in which $v$ appears and $\sigma(s,t)$ is the total number of paths between $s$ and $t$.

Following this idea, the greedy criterion considered in this work consists in performing a breadth first search (BFS) from each node in the network to evaluate the shortest paths between every pair of nodes. During this exploration, the method evaluates how many times the vertex under evaluation appears in the shortest path between two nodes. Then, the candidate vertex with the largest number of appearances in shortest paths is included in the solution. More formally, the greedy function used in this constructive procedure is defined as

$$g(v) = \sum_{s,t \in V \setminus \{v\}} \sigma^S(s,t|v)$$

where $\sigma^S(s,t|v)$ takes value 1 if $v$ is in the shortest path between $s$ and $t$ and 0 otherwise. Notice that this greedy function reduces the complexity with respect to $b(v)$, since it is not necessary to evaluate all the paths between two nodes but evaluate just the shortest path. In particular, the best approach for evaluating the betweenness presents a complexity of $O(m \cdot n + n^2 \log n)$ [24], while the complexity of the proposed method is $O(m \cdot n + n^2)$.

These steps are repeated until including $p$ nodes in the solution. It is worth mentioning that, in each iteration, the nodes that are already included in the solution are not considered for the BFS. This behavior allows the greedy function to be more accurate, selecting the nodes that appears in most of the remaining paths after having removed those edges in which the nodes already in the solution are endpoints.

## 2.2  Shake

The *Shake* method in the VNS methodology is responsible for diversifying the search, allowing BVNS to escape from local optima. In order to do so, the method finds a neighbor solution with respect to the neighborhood under evaluation. Therefore, we first need to define the neighborhood considered in this work. The neighborhood of a given solution $S$ is defined as the set of solutions that can be reached by performing a single move to $S$. We propose the swap move for CNDP, which consists of removing a selected node from the solution and inserting a non-selected node in it. In mathematical terms, the move $Swap(S, u, v)$ is defined as

$$Swap(S, u, v) = S \setminus \{u\} \cup \{v\}$$

Then, the neighborhood $N_s^1(S)$ is conformed by all the solutions that can be reached by removing a node from $S$ and inserting a new one from $V \setminus S$. More formally,

$$N_s^1(S) = \{S' \leftarrow Swap(S, u, v) : \quad \forall u \in S \land \forall v \in V \setminus S\}$$

Without loose of generality, let us define neighborhood $N_s^i(S)$ as the set of solutions that can be reached by performing $i$ swap moves over solution $S$.

Having defined the neighborhood under consideration, the *Shake* procedure randomly selects a neighbor solution in $N_s^k(S)$ to continue the search. Notice that the objective function value of the selected neighbor solution is not relevant for this procedure since it is focused on diversification and, therefore, solutions with larger objective function value (i.e., worse solutions) are accepted.

## 2.3 Improvement

The improvement phase in VNS aims to find a local optimum of a given solution with respect to a certain neighborhood. In the context of VNS, the improvement phase is usually conformed with a local search method, but it can be replaced with a more complex heuristic or even a complete metaheuristic algorithm [29]. Regarding the CNDP, as we aim to design a fast algorithm, we propose the use of a simple but effective local search method.

The first key element in a local search method is the neighborhood to be explored. Given a solution $S$, we propose the exploration of the neighborhood $N_s^1(S)$, defined in Sect. 2.2. There are two main strategies for exploring the neighborhood: Best Improvement and First Improvement. The former evaluates all the solutions in the neighborhood, returning the best solution found in it. The latter, however, explores the neighborhood following a certain order, continuing the search through the first solution that improves the incumbent one. We propose the use of a First Improvement strategy since it does not require to explore the complete neighborhood, thus resulting in a more efficient local search method.

In the First Improvement approach it is important to decide the order in which the neighborhood is explored, since the first improving move is accepted, the search can be skewed due to this ordering. Therefore, we try to include in the solution the most promising nodes before, with the aim of finding improving moves faster. The local search proposed sorts the candidate nodes to enter in the solution in descending order with respect to its degree, since a node with a high degree have more probabilities of being in the path of several pairs of nodes. Then, a swap move between the candidate node and each node already in the solution is performed, restarting the search every time an improvement is found. The method stops when the complete neighborhood is explored without finding any improvement in the incumbent solution.

Notice that a direct implementation of this local search is very computationally demanding, since it requires to evaluate the objective function value in each iteration. However, we propose to maintain the connected components in cache and update them every time a node is removed or inserted in a solution. This idea allow the local search to evaluate the solution with a complexity of $O(c)$, being $c$ the number of connected components, since the number of paths in each connected component can be evaluated in $O(1)$.

### 2.4   Neighborhood Change

The *Neighborhood Change* method is applied at the end of each BVNS iteration, being responsible for selecting the next neighborhood to be explored. Algorithm 2 depicts the pseudocode of the *Neighborhood Change* method considered.

---

**Algorithm 2.** $NeighborhoodChange(S, S'', k)$

1: **if** $CNDP(S'') < CNDP(S)$ **then**                    ▷ Improvement
2:     $S \leftarrow S''$
3:     $k \leftarrow 1$
4: **else**
5:     $k \leftarrow k + 1$
6: **end if**

---

The method receives as input parameters the best solution found $S$, the local optimum found in the improvement phase $S''$, and the current neighborhood $k$. Then, if the candidate solution $S''$ presents a better objective function value than the incumbent one $S$ (step 1), it is updated (step 2), restarting the search from the first neighborhood (step 3). Otherwise, the search continues in the next neighborhood (step 5).

## 3   Computational Results

This section has two main objectives: 1) determine the best values for the parameters of the proposed algorithm and 2) evaluate the performance of the BVNS when comparing it with the best methods found in the state of the art. All the algorithms have been coded in Java 9, and the experiments have been performed in an Intel Core i7 2,1 GHz with 16 Gb of RAM.

The testbed of instances used in this work has been directly derived from the literature [31], which has been also used in the state of the art for the CNDP [28]. It is conformed with a set of 16 instances where the number of vertices ranges from 235 to 5000, while the number of edges ranges from 250 to 4999. The value of $p$ is pre-established for each instance, from 50 in the smallest one to 265 in the largest. We refer the reader to [28] for a detailed description of each instance included in the testbed.

We have performed two types of experiments: preliminary and competitive testing. The former are devoted to evaluate the impact of each element included in the algorithm and to select the best value for the input parameters $k_{max}$ and $T$. The aim of the latter is to evaluate the performance of the presented algorithm when comparing it with the best algorithm found in the state of the art for the CNDP. With the aim of avoiding overfitting, the preliminary experimentation only considers 5 representative instances out of 16.

All the experiments report the following metrics: Avg., the average objective function value; Time (s), the average computing time in seconds; Dev(%) the average deviation with respect to the best solution of the experiment; # Best, times that the algorithm matches the best solution of the experiment.

The first preliminary experiment is designed to evaluate the impact of the local search when coupled with the constructive procedure isolated (i.e., without including it in the VNS framework). To that end, we compare the results obtained by the constructive procedure with the ones resulting from applying the local search procedure to the constructed solution. Figure 2 shows the performance of the constructive procedure when executed either isolated (black bars) or coupled with the local search procedure (gray bars). The vertical axis represent the objective function value, while the percentage over each instance indicates the average deviation of the constructive procedure with respect to the solution improved.

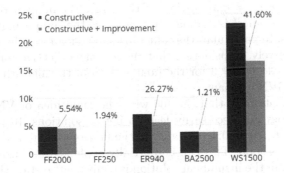

**Fig. 2.** Comparison of constructive procedure isolated and coupled with the local search procedure.

As it can be seen in the figure, even in the smallest instance considered in the preliminary experiment, the local search procedure is able to improve the constructed solution in a 1.94%. Furthermore, the results suggest that the local search method performance grows with the size of the instance. The computing times are not included in the graph since they are all negligible.

The second preliminary experiment is designed to establish the best values for $k_{max}$ and $T$. In this case, a full-factorial experiment is proposed, considering

$k_{max} = \{0.1, 0.2, 0.3, 0.4, 0.5\}$ and $T = \{1, 2, 3, 4, 5\}$. Table 1 shows the average objective function value obtained when considering these $k_{max}$ and $T$ values. The cell color indicates the quality of that combination of parameters, being black the best combination and white the worst one (i.e., the darker, the better).

**Table 1.** Average objective function value of the different configurations for $k_{max}$ and $T$ parameters inside the BVNS algorithm. The darker the cell color, the better the value.

| $T$ \ $k_{max}$ | 0.1 | 0.2 | 0.3 | 0.4 | 0.5 |
|---|---|---|---|---|---|
| 1 | 5594.40 | 5736.60 | 5586.20 | 5700.60 | 5700.40 |
| 2 | 5594.40 | 5623.40 | 5586.00 | 5597.00 | 5566.00 |
| 3 | 5585.80 | 5623.00 | 5520.00 | 5590.80 | 5566.00 |
| 4 | 5544.00 | 5608.20 | 5519.00 | 5589.60 | * |
| 5 | 5531.00 | 5606.00 | 5509.40 | 5589.60 | * |

First of all, we would like to highlight that the cell that contains an asterisk are those combination of parameters that have not been able to report a solution for one or more instances in a time limit of three hour. This behavior indicates that those combinations of parameters are not suitable for being considered in a fast algorithm.

Regarding the average objective function value, the best results are obtained when considering $k_{max} = 0.3$. Obviously, the larger the number of BVNS iterations, the better. Then, it is mandatory to analyze the computing time required for those $T$ values. In particular, the values 3, 4, and 5 for $T$ require from 308, 489, and 667 s, respectively. Considering that the average objective function value is similar, we then select $T = 3$ for the final version of the algorithm, in order to configure a fast BVNS variant.

It is worth mentioning that, in line with the main idea of VNS [17], larger values of $k_{max}$ does not necessarily lead to better solutions. In particular, the results show how $k_{max} = \{0.4, 0.5\}$ are not the best options to set the maximum predefined neighborhood. This can be partially explained because performing a vast perturbation in the incumbent solution is equivalent to start the search from a completely different point, which may not be the best option for the problem under consideration.

Once the best parameters for the proposed algorithm have been selected, the competitive testing evaluates its performance when comparing it with the best previous method found in the state of the art. This method, named GRASP+PR, consists in a Greedy Randomized Adaptive Search Procedure to generate a set of solutions that are later combined using a traditional Path Relinking procedure. In this experiment the full set of 16 instances are considered. Table 2 shows the comparison of the proposed algorithm, BVNS, with the current state of the art method, GRASP+PR.

**Table 2.** Comparison of the BVNS algorithm considering $k_{max} = 0.3$ and $T = 3$ and the best algorithm found in the state of the art, GRASP.

| Algorithm | Avg. | Time (s) | Dev (%) | #Best |
|-----------|------|----------|---------|-------|
| BVNS | 85749.69 | 551.38 | 0.76 | 12 |
| GRASP+PR | 87482.19 | 12774.25 | 9.51 | 10 |

As it can be seen in the table, BVNS is able to reach a smaller objective function value on average (85749.69 versus 87482.19), as well as a smaller deviation with respect to the best solution (0.76% versus 9.51%). The average deviation close to zero indicates that, even in the instances in which BVNS is not able to reach the best solution, it remains very close to it. Furthermore, the number of best solutions found in better in the case of BVNS (12 versus 10). Finally, BVNS is two orders of magnitude faster than GRASP+PR, which confirms that it is able to provide high quality solutions in small computing times, one of the highlights of this work. This results confirms that BVNS emerges as a competitive method for solving the CNDP without requiring high computational efforts.

## 4 Conclusions

This paper proposes a Basic VNS algorithm for generating high quality solutions in short computing times. The initial solution is generated by a novel greedy constructive procedure that leverages the idea of betweenness centrality from Social Network Analysis, while the local improvement consists of a fast first improvement local search procedure based on swap moves which perform an intelligent exploration of the neighborhood. The combination of the constructive procedure and local search method inside a Basic VNS framework results in a competitive algorithm with the best method found in the state of the art, which is able to generate high quality solutions in reasonable computing times.

**Acknowledgements.** This research was funded by "Ministerio de Ciencia, Innovación y Universidades" under grant ref. PGC2018-095322-B-C22, "Comunidad de Madrid" and "Fondos Estructurales" of European Union with grant refs. S2018/TCS-4566, Y2018/EMT-5062.

## References

1. Addis, B., Aringhieri, R., Grosso, A., Hosteins, P.: Hybrid constructive heuristics for the critical node problem. Ann. Oper. Res. **238**(1–2), 637–649 (2016)
2. Addis, B., Di Summa, M., Grosso, A.: Identifying critical nodes in undirected graphs: complexity results and polynomial algorithms for the case of bounded treewidth. Discret. Appl. Math. **161**(16–17), 2349–2360 (2013)
3. Aringhieri, R., Grosso, A., Hosteins, P., Scatamacchia, R.: A general evolutionary framework for different classes of critical node problems. Eng. Appl. Artif. Intell. **55**, 128–145 (2016)

4. Aringhieri, R., Grosso, A., Hosteins, P., Scatamacchia, R.: Local search metaheuristics for the critical node problem. Networks **67**(3), 209–221 (2016)
5. Arulselvan, A.: Network model for disaster management. Ph.D. thesis, University of Florida Gainesville (2009)
6. Arulselvan, A., Commander, C.W., Elefteriadou, L., Pardalos, P.M.: Detecting critical nodes in sparse graphs. Comput. Oper. Res. **36**(7), 2193–2200 (2009)
7. Boginski, V., Commander, C.: Identifying critical nodes in protein-protein interaction networks. In: Clustering Challenges in Biological Networks, February 2009
8. Borgatti, S.P.: Identifying sets of key players in a social network. Comput. Math. Organ. Theory **12**(1), 21–34 (2006). https://doi.org/10.1007/s10588-006-7084-x
9. Bovy, P.H., Thijs, R.: Modelling for transportation systems planning: new approaches and applications. Delft University Press (2002)
10. Commander, C.W., Pardalos, P.M., Ryabchenko, V., Uryasev, S., Zrazhevsky, G.: The wireless network jamming problem. J. Comb. Optim. **14**(4), 481–498 (2007). https://doi.org/10.1007/s10878-007-9071-7
11. Corley, H., Sha, D.Y.: Most vital links and nodes in weighted networks. Oper. Res. Lett. **1**(4), 157–160 (1982)
12. Di Summa, M., Grosso, A., Locatelli, M.: Complexity of the critical node problem over trees. Comput. Oper. Res. **38**(12), 1766–1774 (2011)
13. Di Summa, M., Grosso, A., Locatelli, M.: Branch and cut algorithms for detecting critical nodes in undirected graphs. Comput. Optim. Appl. **53**(3), 649–680 (2012). https://doi.org/10.1007/s10589-012-9458-y
14. Dinh, T.N., Thai, M.T.: Precise structural vulnerability assessment via mathematical programming. In: 2011 - MILCOM 2011 Military Communications Conference, pp. 1351–1356 (2011). https://doi.org/10.1109/MILCOM.2011.6127492
15. Fan, N., Xu, H., Pan, F., Pardalos, P.: Economic analysis of the n - k power grid contingency selection and evaluation by graph algorithms and interdiction methods. Energy Syst. **2**, 313–324 (2011). https://doi.org/10.1007/s12667-011-0038-5
16. Freeman, L.C.: A set of measures of centrality based on betweenness. Sociometry, **40**(1), 35–41 (1977). http://www.jstor.org/stable/3033543
17. Hansen, P., Mladenovic, N., Todosijević, R., Hanafi, S.: Variable neighborhood search: basics and variants. EURO J. Comput. Optim. **5**, 423–454 (2016). https://doi.org/10.1007/s13675-016-0075-x
18. Jhoti, H., Leach, A.R.: Structure-Based Drug Discovery, vol. 1. Springer, Heidelberg (2007). https://doi.org/10.1007/1-4020-4407-0
19. Kempe, D., Kleinberg, J., Tardos, É.: Influential nodes in a diffusion model for social networks. In: Caires, L., Italiano, G.F., Monteiro, L., Palamidessi, C., Yung, M. (eds.) ICALP 2005. LNCS, vol. 3580, pp. 1127–1138. Springer, Heidelberg (2005). https://doi.org/10.1007/11523468_91
20. Lalou, M., Tahraoui, M.A., Kheddouci, H.: The critical node detection problem in networks: a survey. Comput. Sci. Rev. **28**, 92–117 (2018)
21. Li, C.T., Lin, S.D., Shan, M.K.: Finding influential mediators in social networks. In: Proceedings of the 20th International Conference Companion on World Wide Web, WWW 2011, pp. 75–76. Association for Computing Machinery, New York (2011)
22. Liljefors, T., Krogsgaard-Larsen, P., Madsen, U.: Textbook of Drug Design and Discovery. CRC Press, Boco Raton (2002)
23. Mladenović, N., Hansen, P.: Variable neighborhood search. Comput. Oper. Res. **24**(11), 1097–1100 (1997). http://www.sciencedirect.com/science/article/pii/S0305054897000312. https://doi.org/10.1016/S0305-0548(97)00031-2

24. Nasre, M., Pontecorvi, M., Ramachandran, V.: Betweenness centrality – incremental and faster. In: Csuhaj-Varjú, E., Dietzfelbinger, M., Ésik, Z. (eds.) MFCS 2014. LNCS, vol. 8635, pp. 577–588. Springer, Heidelberg (2014). https://doi.org/10.1007/978-3-662-44465-8_49
25. Ovelgönne, M., Kang, C., Sawant, A., Subrahmanian, V.S.: Covertness centrality in networks. In: 2012 IEEE/ACM International Conference on Advances in Social Networks Analysis and Mining, pp. 863–870 (2012). https://doi.org/10.1109/ASONAM.2012.156
26. Petersen, R.R., Rhodes, C.J., Wiil, U.K.: Node removal in criminal networks. In: 2011 European Intelligence and Security Informatics Conference, pp. 360–365 (2011). https://doi.org/10.1109/EISIC.2011.57
27. Pullan, W.: Heuristic identification of critical nodes in sparse real-world graphs. J. Heuristics **21**(5), 577–598 (2015). https://doi.org/10.1007/s10732-015-9290-5
28. Purevsuren, D., Cui, G., Win, N.N.H., Wang, X.: Heuristic algorithm for identifying critical nodes in graphs. Adv. Comput. Sci. Int. J. **5**(3), 1–4 (2016)
29. Sánchez-Oro, J., Martínez-Gavara, A., Laguna, M., Martí, R., Duarte, A.: Variable neighborhood scatter search for the incremental graph drawing problem. Comput. Optim. Appl. **68**(3), 775–797 (2017). https://doi.org/10.1007/s10589-017-9926-5
30. Tomaino, V., Arulselvan, A., Veltri, P., Pardalos, P.M.: Studying connectivity properties in human protein–protein interaction network in cancer pathway. In: Pardalos, P., Xanthopoulos, P., Zervakis, M. (eds.) Data Mining for Biomarker Discovery. SOIA, vol. 65, pp. 187–197. Springer, Boston (2012). https://doi.org/10.1007/978-1-4614-2107-8_10
31. Ventresca, M.: Global search algorithms using a combinatorial unranking-based problem representation for the critical node detection problem. Comput. Oper. Res. **39**(11), 2763–2775 (2012)
32. Ventresca, M., Aleman, D.: A fast greedy algorithm for the critical node detection problem. In: Zhang, Z., Wu, L., Xu, W., Du, D.-Z. (eds.) COCOA 2014. LNCS, vol. 8881, pp. 603–612. Springer, Cham (2014). https://doi.org/10.1007/978-3-319-12691-3_45
33. Veremyev, A., Boginski, V., Pasiliao, E.L.: Exact identification of critical nodes in sparse networks via new compact formulations. Optim. Lett. **8**(4), 1245–1259 (2014). https://doi.org/10.1007/s11590-013-0666-x
34. Vitoriano, B., Ortuño, M.T., Tirado, G., Montero, J.: A multi-criteria optimization model for humanitarian aid distribution. J. Glob. Optim. **51**(2), 189–208 (2011). https://doi.org/10.1007/s10898-010-9603-z
35. Zhou, Y., Hao, J.K.: A fast heuristic algorithm for the critical node problem. In: Proceedings of the Genetic and Evolutionary Computation Conference Companion, pp. 121–122 (2017)

# Variable Neighborhood Descent Branching Applied to the Green Electric Vehicle Routing Problem with Time Window and Mixed Fleet

Thiago M. Stehling[1](✉), Marcone J. Freitas Souza[2](✉),
and Sérgio R. de Souza[1](✉)

[1] Centro Federal de Educação Tecnológica de Minas Gerais (CEFET-MG),
Avenida Amazonas, 7675 – Nova Gameleira, Belo Horizonte, MG, Brazil
thiagostehling@gmail.com, sergio@cefetmg.br
[2] Departamento de Computação, Universidade Federal de Ouro Preto (UFOP),
Campus Universitário – Morro do Cruzeiro, Ouro Preto, MG, Brazil
marcone@ufop.edu.br

**Abstract.** This paper deals with the Green Electric Vehicle Routing Problem with Time Window and Mixed Fleet and presents a Mixed Integer Linear Programming formulation for it. Initially, we applied the CPLEX solver in this formulation. Then, to reduce the computational time, we used Local Branching and Variable Neighborhood Descent Branching (VNDB) methods. We did computational experiments with a simple adaptation of the 100-customers Solomon's benchmark instances. The results showed that the three solution strategies reached the optimal solution. However, the running time of the VNDB is considerably smaller than those required by the other two solution methods. Therefore, this fact proves that the VNDB is the more efficient technique in the tested scenario.

**Keywords:** Green Vehicle Routing Problem · Variable neighborhood search · Variable Neighborhood Descent Branching · Matheuristics · Mathematical programming

## 1 Introduction

In the last few decades, the Logistics and Transport (L&T) systems have been intensively studied by the scientific community because the activities of this sector represent a considerable impact on world economic expenses. A classic problem in L&T is the Vehicle Routing Problem (VRP), which aims to minimize the costs of the routes designed as a path for the vehicles of a given fleet [23]. Variations of the VRP are being proposed over time by the scientific community due to their applicability in real transport situations. For example, the Capacitated Vehicle Routing Problem (CVRP) defines a load capacity for the

© Springer Nature Switzerland AG 2021
N. Mladenovic et al. (Eds.): ICVNS 2021, LNCS 12559, pp. 14–27, 2021.
https://doi.org/10.1007/978-3-030-69625-2_2

fleet of vehicles. The Vehicle Routing Problem with Time Window (VRPTW) extends the CVRP by adding a time window to the depot and the customers. The addition of new constraints to the VRP generates several variations of the problem [4]. However, most of them consider that the fleet of vehicles has an internal-combustion engine, but this fact does not reflect the current scenario in L&T [13].

Evaluating L&T systems from an environmental and sustainable perspective goes beyond the exclusively economic context and makes them more efficient [6]. L&T is responsible for a large percentage of oil consumption in North American and European countries. This fact makes necessary the presence of efficient proposals for systems in this sector. Hence, L&T also is responsible (on a global scale) for a large percentage of the emission of carbon dioxide ($CO_2$) in the Earth's atmosphere. A possible solution for this scenario is to incorporate carbon footprint costs into the problem's model [2]. Adding this feature to the VRP generates a variation known as the Green Vehicle Routing Problem (GVRP), which includes variants with constraints of fuel consumption, pollutant control, and reverse logistics [17]. Another solution that aims to minimize the carbon footprint is the use of alternative vehicles (e.g., electric vehicle, or EV). Including this type of vehicle in the VRP defines a variation known as the Electric Vehicle Routing Problem (EVRP) [13].

The use of EVs is a less polluting transport alternative. However, an efficient L&T system must also deal with energy sources [9]. For example, consider a country in which the principal source of energy is hydroelectric. This source does not emit $CO_2$ (or other pollutants) during the energy generation process. The EVs will use this electricity as fuel, and this type of vehicle emits fewer pollutants than the conventional vehicle (i.e., internal-combustion engine vehicle, or CV). Now, consider a country in which the principal source of energy is thermoelectric. In this case, the burning of coal emits the same amount (or even exceeds it) of $CO_2$ compared to the carbon footprint corresponding to the use of CVs. However, the energy source is not the only barrier to the establishment of EVs in L&T. Their autonomy is also another one. Nowadays, the average driving range of most EVs is 150 miles, and this value can decrease significantly due to cold temperatures [5]. The last barrier identified in the energy scenario is the availability of recharging stations. The absence of these stations is one of the biggest challenges to the success of using EVs in L&T [13].

The use of a mixed fleet of vehicles is a strategy to overcome the barriers coming from the incorporation of EVs in L&T systems. This concept was introduced in the VRP more than thirty years ago. However, the works with a mixed fleet approach in the EVRP are recent [9]. This strategy becomes interesting if we consider the context of technology migration (i.e., from CVs to EVs) [11]. This technological transition is noticeable since some European and Asian countries plan to shut down the CVs' factories in a decade. Moreover, using a mixed fleet provides a balance to the conflicting interests of the problem (i.e., economic versus environmental), and improves routing strategies within the scope of the

EVRP. Hence, it optimizes the proposed models, which result in efficient L&T systems [12].

This paper deals with the Green Electric Vehicle Routing Problem with Time Window and Mixed Fleet (G-EVRPTWMF). This problem does not add the costs of carbon footprint or energy consumption to the problem's objective function, as in the formulation proposed by Figliozzi [6]. We add these features to the problem in the form of constraints. Unlike the variation defined by Yu et al. [24], the G-EVRPTWMF has a mixed fleet of CVs and EVs that have the same load capacity, as in the classic version of the CVRP. The G-EVRPTWMF considers that the EVs leave the depot with a fully charged battery and must return before a recharge is needed, unlike the variants presented by Andelmin and Bartolini [1] and Hiermann et al. [11], who define recharging stations, or by Macrina et al. [18], who establish a partial battery recharging. This feature is quite common in the classic VRPTW formulations, which do not consider filling the CVs' tank at gas stations. Another difference in the G-EVRPTWMF is the definition that the fuel consumption of both types of vehicles is proportional to the travel distance, unlike the model proposed by Goeke and Schneider [9].

The main contributions of this paper are as follows. First, we present the formal definition of the mathematical model of the G-EVRPTWMF, which is a new variation of the problem. Second, we present the implementation of the Variable Neighborhood Descent Branching matheuristic as a solution strategy for the G-EVRPTWMF. This approach shows that it is possible to solve the problem instances in a computational time shorter than that spent by the IBM ILOG CPLEX Optimizer and by the Local Branching matheuristic (i.e., we also applied the proposed mathematical model in both methods).

This paper is structured as follows. Section 2 provides the mathematical formulation of the G-EVRPTWMF, including the parameters and decision variables of the model. Section 3 shows the most recent related works, highlighting the problems and the solution techniques that the authors have proposed. Section 4 explains the methodology applied in this work, as well as the implementation details of the proposed matheuristic. Section 5 defines the testing environment and the benchmark instances used in this work. Also, it shows the computational results and the analysis of them. Finally, Sect. 6 concludes this work.

## 2    Problem Statement

The G-EVRPTWMF is mathematically defined on the complete graph $G = (V, E)$, where $V = \{0, 1, 2, \ldots, N, N + 1\}$ is the vertex set. The vertices 0 and $N + 1$ represent the depot, creating a concept of source vertex and sink vertex, respectively. Also, vertices 1 to $N$ represent the customers. Let $V^* = V\backslash\{0, N + 1\}$, $V^- = V\backslash\{N + 1\}$, and $V^+ = V\backslash\{0\}$. Every route defines a simple path in $G$, which starts at vertex 0 and ends at vertex $N + 1$. Each vertex has a demand $(D)$, a service time $(ST)$, an initial time window $(ITW)$ and a final time window $(FTW)$. Let $D_0 = ST_0 = 0$ and $D_{N+1} = ST_{N+1} = 0$. The range

$[ITW, FTW]$ indicates the time window in which the vehicle must reach the customer's location. In other words, the service cannot be started before $ITW$ or after $FTW$.

The set of edges $E = \{(i, j) : i, j \in V, i \neq j\}$ represents paths between each pair of customers. For example, the edge $(i, j)$ defines the path which starts at $i$ and ends at $j$. Every edge has a cost $(C_{ij})$, which denotes the travel distance from $i$ to $j$. We define the set of CVs $(K_{CV})$, as well as the upper bound $(UB)$ for its number of elements. The emission factor $(\Theta)$ helps to control the carbon footprint of the CVs, which must not exceed the allowed environmental bound $(EB)$. We also define the set of EVs $(K_{EV})$. Each route is traveled by a single CV or EV. The load capacity $(LC)$ is identical for both types of vehicles. In addition to the parameters previously stated, the G-EVRPTWMF has three decision variables. The first and second variables are binary, while the last one assumes real value:

$$X_{ijk} = \begin{cases} 1, & \text{if the CV } k \in K_{CV} \text{ travels the edge } (i, j); \\ 0, & \text{otherwise.} \end{cases}$$

$$Y_{ijk} = \begin{cases} 1, & \text{if the EV } k \in K_{EV} \text{ travels the edge } (i, j); \\ 0, & \text{otherwise.} \end{cases}$$

$Z_{ik} = $ arrival time of vehicle $k$ (unrestrictedly to the type) at the customer $i$

The Mixed Integer Linear Programming (MILP) formulation, which mathematically defines the G-EVRPTWMF, is given by:

$$\text{Minimize} \sum_{k \in K_{CV}} \sum_{i \in V} \sum_{j \in V} C_{ij} X_{ijk} + \sum_{k \in K_{EV}} \sum_{i \in V} \sum_{j \in V} C_{ij} Y_{ijk} \qquad (1)$$

Subject to:

$$\sum_{k \in K_{CV}} \sum_{j \in V^*} X_{ijk} + \sum_{k \in K_{EV}} \sum_{j \in V^*} Y_{ijk} = 1, \forall i \in V^*, i \neq j \qquad (2)$$

$$\sum_{k \in K_{CV}} \sum_{j \in V+} X_{ijk} - X_{jik} = 0, \forall i \in V^*, i \neq j \qquad (3)$$

$$\sum_{k \in K_{EV}} \sum_{j \in V+} Y_{ijk} - Y_{jik} = 0, \forall i \in V^*, i \neq j \qquad (4)$$

$$\sum_{j \in V^*} X_{0jk} \leq 1, \forall k \in K_{CV} \qquad (5)$$

$$\sum_{j \in V^*} Y_{0jk} \leq 1, \forall k \in K_{EV} \qquad (6)$$

$$\sum_{k \in K_{CV}} \sum_{j \in V^*} X_{0jk} \leq UB, \forall k \in K_{CV}, \forall j \in V^* \qquad (7)$$

$$\sum_{i \in V^*} D_i \sum_{j \in V^*} X_{ijk} \leq LC, \forall k \in K_{CV}, i \neq j \qquad (8)$$

$$\sum_{i \in V^*} D_i \sum_{j \in V^*} Y_{ijk} \leq LC, \forall k \in K_{EV}, i \neq j \tag{9}$$

$$Z_{ik} + X_{ijk}(C_{ij} + ST_i) - FTW_0(1 - X_{ijk}) \leq Z_{jk}, \forall i \in V^-, \forall j \in V^+, i \neq j, \forall k \in K_{CV} \tag{10}$$

$$Z_{ik} + Y_{ijk}(C_{ij} + ST_i) - FTW_0(1 - Y_{ijk}) \leq Z_{jk}, \forall i \in V^-, \forall j \in V^+, i \neq j, \forall k \in K_{EV} \tag{11}$$

$$ITW_i \leq Z_{ik} \leq FTW_i, \forall i \in V, \forall k \in K_{CV} \tag{12}$$

$$ITW_i \leq Z_{ik} \leq FTW_i, \forall i \in V, \forall k \in K_{EV} \tag{13}$$

$$\sum_{k \in K_{CV}} \sum_{i \in V} \sum_{j \in V} \Theta C_{ij} X_{ijk} \leq EB \tag{14}$$

$$Z_{ik} \geq 0, \forall i \in V, \forall k \in K_{CV} \tag{15}$$

$$Z_{ik} \geq 0, \forall i \in V, \forall k \in K_{EV} \tag{16}$$

$$X_{ijk} \in \{0, 1\}, \forall i \in V, \forall j \in V, \forall k \in K_{CV} \tag{17}$$

$$Y_{ijk} \in \{0, 1\}, \forall i \in V, \forall j \in V, \forall k \in K_{EV} \tag{18}$$

The formulation of this paper is based on the classic model of the VRPTW (see Toth and Vigo [23]). The objective function minimizes the sum of the total distance traveled by the vehicles used to solve the problem (i.e., CVs and EVs). Constraints (2) ensure that each customer is visited exactly once. Constraints (3) and (4) determine the flow conservation. Constraints (5) and (6) ensure that each vehicle (CV or EV, respectively) is assigned to a single route at maximum.

Constraints (7) insert a cover cut in the problem's formulation (see Kalle-hauge et al. [14]). Constraints (8) and (9) represent the load capacity limitations for the CVs and EVs, respectively. This capacity ($LC$) is identical for both types of vehicles because the mixed fleet is homogeneous. Constraints (10) link the arrival times of the CV $k$ in customers $i$ and $j$ to determine the visit order. These constraints also eliminate any sub-route. Constraints (11) do the same to the EVs. Constraints (12) and (13) ensure that the vehicle $k$ (CV and EV, respectively) will arrive at the customer $i$ (including the depot) within its time window.

Constraints (14) impose an environmental bound on the carbon footprint of the CVs. The emission factor ($\Theta$) is calculated based on the data available in Macrina et al. [18]. In this work, we did not consider the weight variation of the vehicle (i.e., full load versus empty) along the path. Therefore, we measured $\Theta$ as the average of the carbon footprint produced by a full load CV and an empty one. Finally, constraints (15), (16), (17), and (18) define the domains of all variables of the G-EVRPTWMF.

In the context of the vehicles' mixed fleet, we adapted our model based on Goeke and Schneider [9], and the modifications are as follows. First, we introduced a new binary decision variable $(Y_{ijk})$ for the EVs. Second, we add another parcel referring to the sum of the EVs' total traveled distance in the objective function (1). Third, we add another parcel referring to the single visit for the EVs in constraints (2). Finally, we duplicated the constraints (3), (5), (8), (10), (12), (15), and (17).

## 3   Literature Review

Lazić et al. [16] proposed a hybrid heuristic for MILP. This method is a two-level variable neighborhood search scheme. In the first level (i.e., the hard variable fixing strategy), they use a variable neighborhood decomposition search framework. In the second one (i.e., the soft variable fixing strategy), these authors introduce pseudo-cuts as a local search branching method based on a variable neighborhood descent scheme. This combined technique is interesting since it showed better results in some aspects, e.g., average percentage gap, average rank according to objective values and the number of times that the method managed to improve the best known published objective. Bruglieri et al. [3] introduced a MILP formulation for the EVRP with time window constraints. Unlike this work, the authors deal with recharging stations and consider a partial battery recharging. The innovation mentioned by them is that the battery level reached is a decision variable of the optimization process. Bruglieri et al. [3] compare their model with an already existing mathematical formulation that always assumes a full recharge of the EVs. Finally, these authors used a variable neighborhood search branching [10] as a solution method.

Goeke and Schneider [9] used an adaptive large neighborhood search algorithm to solve the EVRP with time window and mixed fleet constraints. Unlike this work, the authors do not deal with the carbon footprint. Also, they add recharging stations to the problem. A local search method enhanced the algorithm. This approach is interesting because it expands the search strategy, diversifying the generated solutions. However, at certain times, this search is reduced to refine these solutions. Hiermann et al. [11] dealt with a similar problem. However, these authors use a heterogeneous fleet, whose EVs differ in load capacity, battery size, and acquisition cost. In addition to the strategy proposed by Goeke and Schneider [9], Hiermann et al. [11] also use an exact method. These authors make a comparison between heuristic and exact results because they introduced a new variation of the problem.

Koç and Karaoglan [15] proposed an exact algorithm for a GVRP variation that uses a fleet of alternative energy-powered vehicles, which run on alternative fuels such as biodiesel, ethanol, hydrogen, natural gas, among others. Unlike this work, the authors define recharging stations and assume that no more than one stop at these stations will be necessary during the travel of any vehicle. Their algorithm applies valid inequalities and uses a heuristic method to improve its bounds. Despite presenting a mathematical formulation that uses fewer variables and constraints, these authors also use a simulated annealing method.

This metaheuristic enhances the initial solution of the algorithm. Andelmin and Bartolini [1] used an exact two-phase algorithm to deal with a similar problem. However, these authors define several gas stations in the geographic space and do not use EVs, despite citing them as another example of alternative energy-powered vehicles. The first phase to the method defines partitions for the problem to form columns of feasible routes, and the second one heuristically lists these routes, but it exactly solves each partition.

Schiffer and Walther [20] proposed an exact algorithm as a solution strategy for the EVRP with time window and partial battery recharging constraints. Unlike this work, the authors solve two problems. At first, the location problem of the recharging stations. Then, the EVRP variation. Schiffer and Walther [20] state that their algorithm can only solve small instances in a reduced computational time. Due to this fact, in Schiffer and Walther [21], they also proposed an adaptive large neighborhood search algorithm to solve large-scale problems. Paz et al. [19] dealt with a similar problem, which has the addition of multiple depots. Also, these authors consider that EVs can be recharged conventionally or by batteries swap. They used an exact algorithm to solve this EVRP variation. The alternative solution proposed by Paz et al. [19] to improve the computational time required by their algorithm was a pre-processing strategy in the constraints that involve the problem's dummy vertices. In this case, this approach is interesting because these authors deal with multiple depots.

Yu et al. [24] used an exact algorithm to solve the GVRP with time window and heterogeneous fleet constraints. Unlike this work, the authors do not use EVs, neither a mixed fleet. Their solution strategy was enhanced by an approximate dynamic programming method and a column generation algorithm. These authors highlight the importance of the integer branch method, which is responsible for obtaining tight upper bounds in a more quickly manner. Foroutan et al. [8] dealt with a GVRP variation that presents a heterogeneous fleet and reverse logistics constraints. The authors solve simultaneously two problems: a scheduling problem, which considers weighted earliness and tardiness costs, and a routing problem. They proposed a non-linear formulation for their problem to solve small instances and heuristic algorithms based on simulated annealing method and genetic algorithm to treat large instances of this GVRP variation.

Macrina et al. [18] proposed an iterated local search algorithm as solution strategy for a similar problem to the G-EVRPTWMF. However, these authors include recharging stations, consider a partial battery recharging, and calculate the carbon footprint and the fuel consumption based on the vehicle load. The authors admit the existence of a set of recharging stations in the geographic space. However, the availability of these stations is not yet a reality (e.g., in emerging countries). As mentioned before, the absence of these stations is one of the biggest challenges to the success of using EVs in L&T [13]. A possible solution would be to solve the location problem of recharging stations and the EVRP together, as is done in Paz et al. and Schiffer and Walther [19–21]. Another solution would be to deal with the fact that EVs will take on smaller routes due to their limited driving autonomy, as is done in this work.

# 4   Variable Neighborhood Descent Branching Framework

At first, we apply the CPLEX solver for the proposed MILP model to validate
it. This optimizer was executed in standard mode. In this paper, we propose an
implementation of the Variable Neighborhood Descent for MILP, named Variable
Neighborhood Descent Branching (VNDB), for solving the G-EVRPTWMF to
reduce the computational time spent by CPLEX. VNDB matheuristic is a deterministic variant of Variable Neighborhood Search Branching (VNSB), which was
introduced by Hansen and Mladenović [10]. Its operation consists of adding linear constraints (i.e., pseudo-cuts) to the problem to systematically change the
neighborhoods. Fischetti and Lodi [7] introduced the concept of a local search
that adds pseudo-cuts into the MILP model in their proposed method known
as Local Branching (LB). A common feature between VNSB and LB is that
they both use a MILP solver (e.g., CPLEX) as a black box for exactly solving
problems in an iterative process [10,16]. Algorithm 1 shows the pseudo-code of
the VNDB matheuristic.

VNDB starts its execution generating a feasible solution $x$ for the original
MILP model (Line 1 of Algorithm 1). Note that $x = (X, Y, Z)$. CPLEX generates
this initial solution (i.e., the first solution found by the solver). Then, the best
solution so far $x^*$ is updated (Line 2) and $\tau = 2$ (Line 3). At this point (Line 4),
VNDB starts its iterative loop. In the Line 5, the pseudo-cuts are added to the
original MILP model to reduce the search space $S$ using a distance function
$\delta(x, \bar{x})$ that generates another space $\bar{S} = S \cap N_\tau(x)$, where:

$$N_\tau(x) = \{\bar{x} \mid \delta(x, \bar{x}) \leq \tau\} \tag{19}$$

In Eq. (19), $N_\tau(x)$ represents the neighborhood structure of the solution $x$.
The distance between the incumbent solution $\bar{x}$ and the previous solution $x$ is
defined by the Hamming distance. In other words, this distance is calculated by
the number of positions in which their binary variables differ, which is limited
by $\tau$. This parameter (i.e., the size of the neighborhood) was adapted to the
scope of the G-EVRPTWMF by the fact that any change of position requires,
at least, two modifications, similarly a swap move. In terms of the problem's
formulation, the pseudo-cut added to the original MILP model is as follows:

$$\sum_{k \in K_{CV}} \sum_{i \in V} \sum_{j \in V} (1 - X_{ijk}) \leq \tau/2 \tag{20}$$

In constraint (20), the $X_{ijk}$ binary decision variable represents the CVs of
the resulting MILP model. However, only the values of $X_{ijk}$ that have been
changed from 1 to 0 are considered. As G-EVRPTWMF has a mixed fleet,
another pseudo-cut must be inserted:

$$\sum_{k \in K_{EV}} \sum_{i \in V} \sum_{j \in V} (1 - Y_{ijk}) \leq \tau/2 \tag{21}$$

In constraint (21), the $Y_{ijk}$ binary decision variable represents the EVs of
the resulting MILP model. The values must be accounted for it in the same

---

**Algorithm 1:** VNDB Algorithm

---

**Input:** original MILP model
**Output:** $x^*$

1  Find a feasible solution $x$ for the original MILP model by using CPLEX;
2  Set $x^* \leftarrow x$;
3  Set $\tau \leftarrow 2$;
4  **while** *stopping condition is not satisfied* **do**
5      Insert the pseudo-cuts $\delta(x, \bar{x}) \leq \tau$ and $\delta(x, \bar{x}) \geq \tau$ into the original MILP model;
6      Find an incumbent solution $\bar{x}$ for the resulting MILP model by using CPLEX;
7      **if** $\bar{x}$ *is an optimal local solution* **then**
8          **if** $\bar{x}$ *is better than* $x$ **then**
9              Set $x \leftarrow \bar{x}$;
10              Remove the pseudo-cuts $\delta(x, \bar{x}) \leq \tau$ and $\delta(x, \bar{x}) \geq \tau$ from the resulting MILP model;
11              **if** $\bar{x}$ *is better than* $x^*$ **then**
12                  Update the bounds of the problem by using the CPLEX information;
13                  Set $x^* \leftarrow \bar{x}$;
14                  Set $\tau \leftarrow 2$;
15              **else**
16                  Set $\tau \leftarrow \tau + 2$;
17          **else**
18              Remove the pseudo-cuts $\delta(x, \bar{x}) \leq \tau$ and $\delta(x, \bar{x}) \geq \tau$ from the resulting MILP model;
19              Set $\tau \leftarrow \tau + 2$;
20      **else**
21          Remove the pseudo-cuts $\delta(x, \bar{x}) \leq \tau$ and $\delta(x, \bar{x}) \geq \tau$ from the resulting MILP model;
22          Set $\tau \leftarrow \tau + 2$;
23      Update the stopping condition by using the CPLEX information;

---

way as in the constraint (20). See Fischetti and Lodi [7] to understand how the asymmetric shape of these pseudo-cuts was defined.

In Line 6 of Algorithm 1, CPLEX is called again, but to solve the sub-problem generated after adding the pseudo-cuts. If an optimal local solution was found (Line 7), then VNDB proceeds its operation. Otherwise, CPLEX reports the proven infeasibility of the space $\bar{S}$. Therefore, VNDB removes the previously inserted pseudo-cuts (Line 21) and updates $\tau$ (Line 22). If $\bar{x}$ improves the objective function value of $x$ (Line 8), then VNDB proceeds its operation. Otherwise, CPLEX reports the optimal local solution found within the space $\bar{S}$. However, $\bar{x}$ is not an improvement solution. Therefore, VNDB removes the previously inserted pseudo-cuts (Line 18) and updates $\tau$ (Line 19).

At this point (Line 9 of Algorithm 1), $x$ is updated by the improvement solution $\bar{x}$. In Line 10, VNDB removes the previously inserted pseudo-cuts. If $\bar{x}$ is the best solution so far (Line 11), then the problem bounds are updated (Line 12), $x^*$ also is updated (Line 13), and the search returns to the initial neighborhood (Line 14). Otherwise, CPLEX reports the optimal local solution found within the space $\bar{S}$. However, $\bar{x}$ is not the best solution so far. Thus, the search proceeds to the next neighborhood (Line 16). As mentioned before, VNDB is a deterministic variant of VNS for MILP. It is true because the maximum time per node in the search tree was not stipulated. Hence, CPLEX can report only two situations, i.e., optimal local solution found in $\bar{S}$ or proven infeasibility of this space. Furthermore, the diversification mechanism was not used in this work (see more details in Fischetti and Lodi [7]).

In each iteration of the algorithm, the stopping criterion is updated (Line 23 of Algorithm 1). VNDB ends its operation if the maximum execution time is exceeded or if the optimality gap is reached. This total execution time limit was set based on the running time of obtained by the CPLEX solver. Besides, the optimality gap is the positive difference between the bounds obtained in the last iteration of the method. This interval indicates that such a difference may not be exactly zero, but a value very close to zero.

## 5    Computational Experiments

The mathematical model and the proposed algorithms were coded in C++ by using the ILOG Concert Technology. The MILP model was solved using the IBM CPLEX Optimizer 12.9, with its parameters assuming the default values. All experiments were conducted on an Intel Core i7-4510U 3.1GHz with 16GB RAM, running Ubuntu 18.04.4. Instances of Solomon [22] involving 100 customers were used to test the solution methods. They are available at https://www.sintef.no/projectweb/top/vrptw/solomon-benchmark/100-customers. Altogether, there are 56 instances divided into 6 classes: C1, C2, R1, R2, RC1, and RC2. Customers are distributed geographically in a grouped manner in classes C1 and C2, randomly in R1 and R2, and classes RC1 and RC2 present a mixture of these two characteristics. Moreover, in classes C1, R1, and RC1, customers' time windows are short. Whereas, in classes C2, R2, and RC2, customers' time windows are wide. These instances were proposed for VRPTW. However, for treating G-EVRPTWMF, we consider that the number of EVs is the same as the number of CVs. In other words, if there are 25 CVs in a particular instance, there will also be the same number of EVs.

Table 1 shows the average results obtained by CPLEX, LB, and VNDB, which were grouped by class. Column ($\eta$) indicates the number of instances that each class has. Column ($\eta^*$) represents the number of instances solved optimally by the method in question. Column (CVs + EVs) shows the number of CVs and EVs (i.e., average number grouped by class) used in the solution, respectively. For example, instances of class C1 use 7 CVs and 8 EVs in their solution, resulting in a total of 15 used vehicles. Column (Distance) shows the average traveled distance, and Column (Time) informs the average computational time consumed by the methods, in seconds, to solve the specific class of the instances. The bottom row of the table totals the average values for each column.

**Table 1.** Average results per class of the 100-customers benchmark instances.

| Class | $\eta$ | CPLEX | | | | LB | | | | VNDB | | | |
|---|---|---|---|---|---|---|---|---|---|---|---|---|---|
| | | $\eta^*$ | CVs + EVs | Distance | Time | $\eta^*$ | CVs + EVs | Distance | Time | $\eta^*$ | CVs + EVs | Distance | Time |
| C1 | 9 | 9 | 7 + 8 | 1,127.1 | 4,153.5 | 9 | 7 + 8 | 1,127.1 | 2,146.1 | 9 | 7 + 8 | 1,127.1 | 680.1 |
| C2 | 8 | 8 | 2 + 5 | 1,037.7 | 4,247.3 | 8 | 2 + 5 | 1,037.7 | 2,241.2 | 8 | 2 + 5 | 1,037.7 | 781.7 |
| R1 | 12 | 12 | 8.9 + 8.9 | 2,507.9 | 4,441.4 | 12 | 8.9 + 8.9 | 2,507.9 | 2,453.8 | 12 | 8.9 + 8.9 | 2,507.9 | 925.4 |
| R2 | 11 | 11 | 3.5 + 7.4 | 2,089.3 | 4,542.8 | 11 | 3.5 + 7.4 | 2,089.3 | 2,557.5 | 11 | 3.5 + 7.4 | 2,089.3 | 1,028.5 |
| RC1 | 8 | 8 | 8.4 + 9.3 | 2,406.7 | 4,744.5 | 8 | 8.4 + 9.3 | 2,406.7 | 2,767.5 | 8 | 8.4 + 9.3 | 2,406.7 | 1,173.3 |
| RC2 | 8 | 8 | 4.5 + 8.5 | 1,967.3 | 4,849.3 | 8 | 4.5 + 8.5 | 1,967.3 | 2,857.3 | 8 | 4.5 + 8.5 | 1,967.3 | 1,269.3 |
| **Total** | 56 | 56 | 34.3 + 47.1 | 11,136.0 | 26,978.8 | 56 | 34.3 + 47.1 | 11,136.0 | 15,023.4 | 56 | 34.3 + 47.1 | 11,136.0 | 5,858.3 |

According to Table 1, the three methods achieved optimal results on all 56 instances of Solomon [22]. However, the difference between them occurs regarding the computational time consumed. Evaluating the total average time, VNDB showed a reduction of 2.5 h concerning the average time consumed by LB and almost 6 h to the average time consumed by CPLEX to solve G-EVRPTWMF. Also, in each class, the average time was reduced (practically) by half in both comparisons, i.e., from CPLEX with LB and from LB with VNDB. One explanation for this behavior, especially between LB and VNDB, is how the neighborhood structure varies in each iteration.

Let $\tau = 2$. The pseudo-cut inserted by LB is the type $\delta(x, \bar{x}) \le \tau$. So, the algorithm will explore all solutions that differ by one or two positions away (since $\tau$ is an integer and positive value). However, as stated before, in the scope of the G-EVRPTWMF, any change of position requires, at least, two modifications to be performed, as a swap move. This fact determines that the neighborhood $N_1(x)$ does not generate any feasible solution. This obstacle is avoided by VNDB. This is done by inserting two pseudo-cuts, which are $\delta(x, \bar{x}) \le \tau$ and $\delta(x, \bar{x}) \ge \tau$. Together, these pseudo-cuts define that only the neighborhood $N_2(x)$ will be explored. This simple modification prevents VNDB from wasting time in odd neighborhoods, which do not generate feasible solutions for the G-EVRPTWMF. Also, LB performs a search in neighborhoods in an exclusively increasing way. On the other hand, VNDB returns to the initial neighborhood whenever the best current solution is updated, which generates a more efficient search through variable neighborhoods.

Figure 1 shows the convergence curve of matheuristics in the six classes of instances. In this figure, one can see the advantage of VNDB over CPLEX and LB. The quality of the solution produced by VNDB is always better than that of the other methods in the same processing time, in seconds.

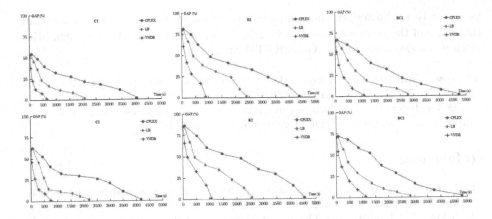

**Fig. 1.** Convergence curve of the methods per class.

In Fig. 1, it is possible to observe the progression of the optimality gap as a function of the average time consumed by the three methods in the 6 classes of the problem. The final processing times, computed when the stopping criterion for each method is reached, can be seen in Table 1. For example, the time spent by the 9 instances of class C1 was 4153.3 (CPLEX), 2146.1 (LB), and 680.1 (VNDB) seconds. Regarding the difficulty of generating good solutions for G-EVRPTWMF, the classes R1 and R2 are the most complexes (note that the initial gaps are the highest). Regarding the time spent to find better solutions to the problem, the classes RC1 and RC2 are the most difficult (note that the average times consumed are the longest in all methods). Finally, it is also possible to verify that the search strategies in variable neighborhoods of VNDB proved to be effective since the matheuristic convergence curve reduces considerably before the first 500 s.

## 6   Concluding Remarks

In this paper, we present the G-EVRPTWMF, which is a problem that combines two conflicting elements in L&T (i.e., the economic and the environmental one). Also, we propose a MILP formulation for it. Initially, we applied the CPLEX solver in this formulation. Then, to reduce the computational time, we used LB and VNDB methods. We explain in detail all the VNDB's adaptations to deal with the mixed fleet (i.e., a fleet that has CVs and EVs). We did computational experiments with a simple adaptation of the 100-customers Solomon's instances. The results showed that the three solution strategies reached the optimal solution. However, the running time of the VNDB is considerably smaller than those required by the other two solution methods. Therefore, this fact proves that the VNDB is the more efficient technique to solve the G-EVRPTWMF. Future research will concentrate on the following issues. First, within the scope of the VNDB, it is possible to implement a version of the method with a more heuristic behavior to try to reduce the computational time even further. Second, within the scope of the problem, it would be interesting to see how other matheuristics work when dealing with the G-EVRPTWMF.

**Acknowledgments.** The authors thank the Brazilian agencies FAPEMIG (grant PPM-CEX 676/17), CNPq (grant 303266/2019-8) and CAPES, as well as the Federal Center for Technological Education of Minas Gerais (CEFET-MG) and the Federal University of Ouro Preto (UFOP) for supporting this study.

# References

1. Andelmin, J., Bartolini, E.: An exact algorithm for the green vehicle routing problem. Transp. Sci. **51**(4), 1288–1303 (2017)
2. Bektaş, T., Laporte, G.: The pollution-routing problem. Transp. Res. Part B: Methodol. **45**(8), 1232–1250 (2011)
3. Bruglieri, M., Pezzella, F., Pisacane, O., Suraci, S., et al.: A variable neighborhood search branching for the electric vehicle routing problem with time windows. Electron. Notes Discret. Math. **47**(15), 221–228 (2015)
4. Caceres-Cruz, J., Arias, P., Guimarans, D., Riera, D., Juan, A.A.: Rich vehicle routing problem: survey. ACM Comput. Surv. (CSUR) **47**(2), 32 (2015)
5. Feng, W., Figliozzi, M.: An economic and technological analysis of the key factors affecting the competitiveness of electric commercial vehicles: a case study from the USA market. Transp. Res. Part C: Emerg. Technol. **26**, 135–145 (2013)
6. Figliozzi, M.: Vehicle routing problem for emissions minimization. Transp. Res. Rec. **2197**(1), 1–7 (2010)
7. Fischetti, M., Lodi, A.: Local branching. Math. Program. **98**(1–3), 23–47 (2003)
8. Foroutan, R.A., Rezaeian, J., Mahdavi, I.: Green vehicle routing and scheduling problem with heterogeneous fleet including reverse logistics in the form of collecting returned goods. Appl. Soft Comput. **94**(1568–4946), 106462 (2020)
9. Goeke, D., Schneider, M.: Routing a mixed fleet of electric and conventional vehicles. Eur. J. Oper. Res. **245**(1), 81–99 (2015)
10. Hansen, P., Mladenović, N., Urošević, D.: Variable neighborhood search and local branching. Comput. Oper. Res. **33**(10), 3034–3045 (2006)
11. Hiermann, G., Puchinger, J., Ropke, S., Hartl, R.F.: The electric fleet size and mix vehicle routing problem with time windows and recharging stations. Eur. J. Oper. Res. **252**(3), 995–1018 (2016)
12. Hung, Y.C., Michailidis, G.: Optimal routing for electric vehicle service systems. Eur. J. Oper. Res. **247**(2), 515–524 (2015)
13. Juan, A., Mendez, C., Faulin, J., De Armas, J., Grasman, S.: Electric vehicles in logistics and transportation: a survey on emerging environmental, strategic, and operational challenges. Energies **9**(2), 86 (2016)
14. Kallehauge, B., Larsen, J., Madsen, O.B., Solomon, M.M.: Vehicle routing problem with time windows. In: Desaulniers, G., Desrosiers, J., Solomon, M.M. (eds.) Column Generation, pp. 67–98. Springer, Boston (2005). https://doi.org/10.1007/0-387-25486-2_3
15. Koç, Ç., Karaoglan, I.: The green vehicle routing problem: a heuristic based exact solution approach. Appl. Soft Comput. **39**, 154–164 (2016)
16. Lazić, J., Hanafi, S., Mladenović, N., Urošević, D.: Variable neighbourhood decomposition search for 0–1 mixed integer programs. Comput. Oper. Res. **37**(6), 1055–1067 (2010)

17. Lin, C., Choy, K.L., Ho, G.T., Chung, S.H., Lam, H.: Survey of green vehicle routing problem: past and future trends. Expert Syst. Appl. **41**(4), 1118–1138 (2014)
18. Macrina, G., Pugliese, L.D.P., Guerriero, F., Laporte, G.: The green mixed fleet vehicle routing problem with partial battery recharging and time windows. Comput. Oper. Res. **101**, 183–199 (2019)
19. Paz, J., Granada-Echeverri, M., Escobar, J.: The multi-depot electric vehicle location routing problem with time windows. Int. J. Ind. Eng. Comput. **9**(1), 123–136 (2018)
20. Schiffer, M., Walther, G.: The electric location routing problem with time windows and partial recharging. Eur. J. Oper. Res. **260**(3), 995–1013 (2017)
21. Schiffer, M., Walther, G.: An adaptive large neighborhood search for the location-routing problem with intra-route facilities. Transp. Sci. **52**(2), 331–352 (2018)
22. Solomon, M.M.: Algorithms for the vehicle routing and scheduling problems with time window constraints. Oper. Res. **35**(2), 254–265 (1987)
23. Toth, P., Vigo, D.: Vehicle Routing: Problems, Methods, and Applications. SIAM (2014)
24. Yu, Y., Wang, S., Wang, J., Huang, M.: A branch-and-price algorithm for the heterogeneous fleet green vehicle routing problem with time windows. Transp. Res. Part B: Methodol. **122**, 511–527 (2019)

# A Variable Neighborhood Heuristic for Facility Locations in Fog Computing

Thiago Alves de Queiroz[1]([✉]) [iD], Claudia Canali[2] [iD], Manuel Iori[3] [iD], and Riccardo Lancellotti[2] [iD]

[1] Institute of Mathematics and Technology, Federal University of Catalão, Catalão 75704-020, GO, Brazil
taq@ufg.br
[2] Department of Engineering "Enzo Ferrari", University of Modena and Reggio Emilia, 41125 Modena, Italy
{claudia.canali,riccardo.lancellotti}@unimore.it
[3] Department of Sciences and Methods for Engineering, University of Modena and Reggio Emilia, 42122 Reggio Emilia, Italy
manuel.iori@unimore.it

**Abstract.** The current trend of the modern smart cities applications towards a continuous increase in the volume of produced data and the concurrent need for low and predictable latency in the response time has motivated the shift from a cloud to a fog computing approach. A fog computing architecture is likely to represent a preferable solution to reduce the application latency and the risk of network congestion by decreasing the volume of data transferred to cloud data centers. However, the design of a fog infrastructure opens new issues concerning not only how to allocate the data flow coming from sensors to fog nodes and from there to cloud data centers, but also the choice of the number and the location of the fog nodes to be activated among a list of potential candidates. We model this facility location issue through a multi-objective optimization problem. We propose a heuristic based on the variable neighborhood search, where neighborhood structures are based on swap and move operations. The proposed method is tested in a wide range of scenarios, considering a smart city application's realistic setup with geographically distributed sensors. The experimental evaluation shows that our method can achieve stable and better performance concerning other literature approaches, supporting the given application.

**Keywords:** Smart cities · Fog networking · Facility location problem

## 1 Introduction

Smart city applications that require the processing of huge volumes of data produced by geographically distributed sensors represent a typical scenario where fog computing is likely to be a winning approach. Its potential has been demonstrated, indeed, by several studies in literature [6,16–18]. The main characteristic

© Springer Nature Switzerland AG 2021
N. Mladenovic et al. (Eds.): ICVNS 2021, LNCS 12559, pp. 28–42, 2021.
https://doi.org/10.1007/978-3-030-69625-2_3

of fog computing is the ability to push on the network edge functions such as data filtering and aggregation [17, 18], with a twofold advantage. First, it reduces the data volume reaching the cloud data center, thus avoiding the risk of poor performance due to high network utilization and reducing the non-negligible economic costs related to the cloud pricing model. Second, the fog layer located on the network edge can guarantee a fast response to latency-sensitive applications (e.g., traffic monitoring and support for autonomous driving) that cannot accept delays in the order of hundreds of milliseconds due to the potentially high round-trip-time latency with the cloud data center.

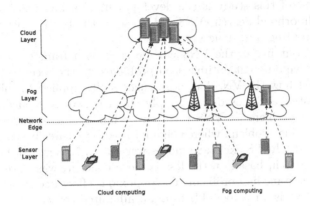

**Fig. 1.** Cloud and fog infrastructures.

In Fig. 1, we compare fog and cloud approaches. In the cloud architecture (left part of the figure), a set of sensors sends data directly to the cloud data center for processing. In the fog case (right part of the figure), a layer of fog nodes is placed on the network edge to host pre-processing, filtering, and aggregation tasks.

The introduction of the intermediate layer of fog nodes represents an additional degree of freedom that arises new issues for the overall infrastructure design. In particular, many studies [6, 19] consider just the fog to cloud communication, adopting a naive approach in the allocation (i.e., mapping) of data flows coming from the sensors over the fog layer, assuming that every sensor sends data to the nearest fog node. On the other hand, recent studies demonstrated that optimized data flow allocation could provide a significant advantage [3]. However, even when some optimization is performed in the sensors-to-fog mapping, no attention has been devoted to minimizing the number of fog nodes required to satisfy the Service Level Agreement (SLA) to reduce the costs and the energy consumption related to the management of the fog infrastructure.

In the operational research field, this issue is named facility location problem [7, 8] and concerns the identification of facilities, so the costs incurred from allocating customers to the selected facilities are minimized, and represents nowadays a very active field of research. A recent survey on service facility location problems can be found in [5]. While this aspect has been widely explored at the level of managing resources in a cloud data center [1,12], it has not been considered so far in the fog computing field.

In this paper, we formalize the facility location problem through an optimization model aiming to map sensors data flow over the fog layer with a two-fold objective: minimize the number of turned on fog nodes while guaranteeing the respect of a service level agreement on response time; and, minimize the response time for the given number of selected fog nodes. In the model, we consider both network delays and processing time at the level of fog nodes. Furthermore, a qualifying point of this study is the development of a heuristic, based on the Variable Neighborhood Search (VNS) [9], to solve the facility location problem over a fog computing infrastructure.

As pointed out in [9], the VNS can be seen as a framework for building heuristics to solve different optimization problems. In the recent survey of [15], it was discussed how the VNS had been successfully applied to solve problems in reverse logistics and closed-loop supply chain networks. Concerning facility location related problems, a VNS with path-relinking was proposed in [20] for the location routing problem, where neighborhood structures based on insertion, swap, 2-opt, and CROSS-exchange moves were used. Recently, a basic variable neighborhood search, based on the less is more concept, was developed in [13] for an obnoxious p-median problem. The obnoxious effect occurs when a facility should be located as far as possible from an inhabited center.

We evaluate the proposed VNS in the realistic scenario of a smart city application with geo-referenced sensors collecting data for traffic monitoring in the city of Modena in Italy. The VNS is compared with the optimization model solved by a state-of-the-art solver in terms of its capability of reducing costs and response times. Moreover, a sensitivity analysis is performed concerning the infrastructure size. The experimental results demonstrate that our proposal can outperform the alternatives with stable results concerning a wide range of scenarios.

The rest of the paper is organized as follows. Section 2 presents the theoretical modeling for the considered problem. Section 3 presents the VNS to solve the facility location problem in fog computing infrastructures. Section 4 presents the experimental setup and the considered scenarios, providing a thorough evaluation of the proposed model against the alternatives. Finally, Sect. 5 gives some concluding remarks and outlines some future work direction.

## 2    Problem Definition

In this section, we formalize the location-allocation problem in a fog architecture as a multi-objective optimization problem that aims to minimize two aspects: 1) the delay in the transit of the data from sensors to fog nodes to the cloud data center; 2) the cost associated with the number of fog nodes turned on.

To model this problem, let us assume a stationary scenario with a set $S$ of geographically distributed sensors producing data at a steady rate: we denote as $\lambda_i$ the frequency of sensor $i$. The data are sent to an intermediate layer of a set $\mathcal{F}$ of fog nodes. These nodes can perform operations on data such as filtering and aggregation or specific analysis, such as anomaly detection with low latency. We denote as $\mu_j$ the processing rate of the fog node $j$, and as $\delta_{ij}$ the delay from sensor $i$ to fog node $j$. The model also includes a set $\mathcal{C}$ of cloud data centers that receive data from the fog nodes, where $\delta_{jk}$ represents the delay from fog node $j$ to cloud data center $k$. To formalize the problem, we use the following binary decision variables: a) $x_{ij}$, indicating whether sensor $i$ sends data to fog node $j$; b) $y_{jk}$, indicating whether fog node $j$ sends data to cloud data center $k$; c) $E_j$, defining whether the fog node located at position $j$ is turned on and available to process data from sensors. The main symbols of the model are summarized in Table 1.

**Table 1.** Notation and parameters for the proposed model.

| Model parameters | |
|---|---|
| $\mathcal{S}$ | Set of sensors |
| $\mathcal{F}$ | Set of fog nodes |
| $\mathcal{C}$ | Set of cloud data centers |
| $\lambda_i$ | Outgoing data rate from sensor $i$ |
| $\lambda_j$ | Incoming data rate at fog node $j$ |
| $1/\mu_j$ | Processing time at fog node $j$ |
| $\delta_{ij}$ | Communication latency between sensor $i$ and fog node $j$ |
| $\delta_{jk}$ | Communication latency between fog $j$ and cloud $k$ |
| $c_j$ | Cost for locating a fog node at position $j$ (or for keeping the fog node turned on) |
| Model indices | |
| $i$ | Index for a sensor |
| $j$ | Index for a fog node |
| $k$ | Index for a cloud data center |
| Decision variables | |
| $E_j$ | Location of fog node $j$ |
| $x_{ij}$ | Allocation of sensor $i$ to fog node $j$ |
| $y_{jk}$ | Allocation of fog node $j$ to cloud data center $k$ |

For the problem of sensors allocation on the fog nodes, introduced in [3], we consider the application average response time $T_R$, which depends on three components (Eq. (1)): the network delay due to the sensor-to-fog latency $T_{netSF}$ (Eq. (2)), the network delay due to the fog-to-cloud latency $T_{netFC}$ (Eq. (3)), and the processing time on the fog nodes $T_{proc}$ (Eq. (4)).

$$T_R = T_{netSF} + T_{netFC} + T_{proc} \tag{1}$$

$$T_{netSF} = \frac{1}{\sum_{i \in S} \lambda_i} \sum_{i \in S} \sum_{j \in \mathcal{F}} \lambda_i x_{ij} \delta_{ij} \tag{2}$$

$$T_{netFC} = \frac{1}{\sum_{j \in \mathcal{F}} \lambda_j} \sum_{j \in \mathcal{F}} \sum_{k \in \mathcal{C}} \lambda_j y_{jk} \delta_{jk} \tag{3}$$

$$T_{proc} = \frac{1}{\sum_{j \in \mathcal{F}} \lambda_j} \sum_{j \in \mathcal{F}} \lambda_j \frac{1}{\mu_j - \lambda_j} \tag{4}$$

In the components $T_{netSF}$ and $T_{netFC}$, each delay is weighted by the amount of traffic exchanged on the corresponding link, which is $\lambda_i$ for $T_{netSF}$ and $\lambda_j$ in $T_{netFC}$. The incoming data rate on each fog node $\lambda_j$, indeed, can be defined as the sum of the data rates of the sensors allocated to that node:

$$\lambda_j = \sum_{i \in S} x_{ij} \lambda_i, \quad \forall j \in \mathcal{F} \tag{5}$$

The component concerning the processing time $T_{proc}$ is modeled using the queuing theory considering an M/G/1 system and is consistent with other results in literature [1,4]. Specifically, the generic fog node $j$ is characterized by an average processing time $1/\mu_j$ and receives an incoming stream of jobs with frequency $\lambda_j$ (where $\lambda_j$ is defined as in Eq. (5)). It is worth mentioning that we do not consider the cloud layer's details in our problem modeling, such as the computation time at the cloud data center level, as this aspect has been widely covered in literature [2].

Finally, we consider that a fixed cost $c_j$ is associated with the fog node $j$ if it is turned on to model the overall cost due to the fog node activation. In our model, we do no consider other constraints, such as the amount of memory required by the fog nodes' application. Such additional constraints can be straightforwardly added to the model. However, in our experience, the most critical constraint for fog infrastructures deployment is the processing power rather than the available memory. Hence, we opted for a more streamlined model. The complete model for the fog node location-allocation problem may be defined as follows.

Minimize:

$$C = \sum_{j \in \mathcal{F}} c_j E_j \tag{6}$$

$$T_R = T_{netSF} + T_{netFC} + T_{proc} \tag{7}$$

Subject to:

$$T_R \leq T_{SLA} \tag{8}$$

$$\lambda_j < E_j \mu_j, \quad \forall j \in \mathcal{F} \tag{9}$$

$$\sum_{j \in \mathcal{F}} x_{ij} = 1, \quad \forall i \in S \tag{10}$$

$$\sum_{k \in \mathcal{C}} y_{jk} = E_j, \quad \forall j \in \mathcal{F} \tag{11}$$

$$E_j \in \{0,1\}, \quad \forall j \in \mathcal{F} \tag{12}$$

$$x_{ij} \in \{0,1\}, \quad \forall i \in \mathcal{S}, j \in \mathcal{F} \tag{13}$$

$$y_{jk} \in \{0,1\}, \quad \forall j \in \mathcal{F}, k \in \mathcal{C} \tag{14}$$

The two objective functions, (6) and (7), are related, respectively, to the minimization of: **cost**, which is associated with the number of fog nodes turned on; and, **latency**, which is the delay in sensor to fog to cloud data transit expressed through the function introduced as $T_R$ in (1). The second objective is subordinated to the first one, meaning that we aim to minimize (7) as long as the improvement for this objective function does not introduce an increase for (6).

The model includes several constraints. Constraint (8) introduces a limit for the average response time that should not exceed a *Service Level Agreement* (SLA), which is typically defined as a multiple of the average response time $1/\mu$ [1]. We also introduce a term due to the network delays in a distributed architecture (non-negligible) that we consider depending on the average network delays $\delta$. The SLA limit in (15) can be formalized as follows, with $K$ defined depending on the network requirements:

$$T_{SLA} = \frac{K}{\mu} + 2\delta \tag{15}$$

Constraints (9) ensure that no overload occurs on each fog node, imposing that the incoming data flow does not exceed the processing rate. For a node that is powered down, no processing must occur. Constraints (10) guarantee for each sensor that one fog node processes its data, while constraints (11) ensure for each fog node that exactly one cloud data center receives its processed data. Constraints (12), (13) and (14) describe the domain of the decision variables.

## 3 Variable Neighborhood Heuristic

A *Variable Neighborhood Search* (VNS) is proposed to solve the facility location problem that arises in fog computing infrastructures. The VNS methodology was initially proposed in [14]. It has a systematical change of neighborhoods to look for a globally optimal solution concerning all neighborhoods.

The VNS for the problem has the following main steps: to create an initial solution, which is the current solution; to obtain a neighbor solution of the current solution by using a neighborhood operator (shake-phase); to perform a local search on the neighbor solution (local-search-phase); and, to accept the solution of the local search if it is better than the current solution, besides updating the current solution. If the current solution is updated, the VNS restarts from the first neighborhood; otherwise, it proceeds to the next neighborhood, repeating the steps above until reaching the last neighborhood [9,10].

In the proposed VNS, a solution $x$ is coded as a vector of lists of integers. Each position of the vector represents a fog node and contains a list of integers with the sensors. There is another integer indicating which cloud data center is serving the fog node. A position with an empty list of integers represents a fog node turned off. Notice the vector has size $|\mathcal{F}|$, and each list of integers has size at most $|\mathcal{S}|$. Moreover, the lists of integers have a null intersection since a sensor is serviced by exactly one fog node.

The initial solution of the VNS is created as follows. For each fog node, we select the closest cloud data center to allocate it. For each sensor, we choose the closest fog node to assign it. If a fog node has already reached the $T_{SLA}$, no other sensor can be allocated to it. It means the second closest fog node is used to allocate the sensor, and so on until all sensors are allocated to fog nodes. Once the initial solution is created, it has its two objectives calculated: (i) the cost associated with the number of fog nodes turned on; and (ii) the delay in sensor to fog to cloud transit of data, where the first objective is used to guide the VNS. We do not accept solutions that violate constraints (8) to (11) in the optimization process. A solution $x'$ is better than another $x''$ if (i) the first objective of $x'$ is smaller than that of $x''$, or if (ii) the two first objectives are equal but the second objective of $x'$ is smaller than the second of $x''$.

As the VNS iterates through $K$ neighborhood structures, we propose five structures based on swap and move operations. In particular:

- $N_1$: select (randomly) a fog node $f_1$, the farthest sensor $s_1$ allocated to $f_1$, the fog $f_2$ that is the closest to $s_1$, and the sensor $s_2$ allocated to $f_2$ that is the closest to $f_1$. Hence, swap $s_1$ and $s_2$ from their respective fog nodes, if this new solution is feasible.
- $N_2$: Let $\mathcal{F}_{on}$ be the set of fog nodes turned on. Compute the load of each fog node $j \in \mathcal{F}_{on}$ as $r_j = \lambda_j/\mu_j$ and, then, the average load of fog nodes turned on as: $\bar{r} = \left(\sum_{j \in \mathcal{F}_{on}} r_j\right)/|\mathcal{F}|$. Select (randomly) $f_1 \in \mathcal{F}_{on}$ whose load $r_1 > \bar{r}$. If one exists, select the farthest sensor $s_1$ allocated to $f_1$. Then, select the fog node $f_2 \in \mathcal{F}_{on}$ with the lowest load $r_2$ and closest to $s_1$. Remove $s_1$ from $f_1$ and move it to $f_2$, if this new solution is feasible.
- $N_3$: Let $\mathcal{F}_{on}$ be the set of fog nodes turned on. Select (randomly) a fog node $f_1 \in \mathcal{F}_{on}$. Compute the average load with all sensors and fog nodes turned on, except $f_1$, as: $\tilde{r} = \left(\sum_{i \in \mathcal{S}} \lambda_i\right) / \left(\sum_{j \in \mathcal{F}_{on} \setminus \{f_1\}} \mu_j\right)$. If $\tilde{r} < 1$, then for each sensor $s_1$ allocated to $f_1$, remove $s_1$ from $f_1$ and move it to the closest fog node in $\mathcal{F}_{on} \setminus \{f_1\}$, if this new solution is feasible.
- $N_4$: Let $\mathcal{F}_{on}$ be the set of fog nodes turned on. Let $\mathcal{F}_{off}$ be the set of fog nodes turned off. If $\mathcal{F}_{off}$ is not empty, select (randomly) a fog node $f_1 \in \mathcal{F}_{off}$. Select the fog node $f_2 \in \mathcal{F}_{on}$ whose average response time is the highest one. Remove all sensors from $f_2$ and move them to $f_1$, then turning off $f_2$, if this new solution is feasible.
- $N_5$: if the number of available cloud data centers is greater than one, select (randomly) a fog node turned on and allocate it to the closest cloud data center, if this new solution is feasible.

Regarding the shake-phase of the VNS, the selection of fog nodes in the neighborhood structures occurs randomly if the contrary is not imposed. On the other hand, in the local-search-phase, we use a variable neighborhood descent based procedure, in which the solution from the shake-phase is passed as the input parameter [9]. In this procedure, two neighborhood structures are used. It consists of performing all possible (i) allocations of sensors in fog nodes and (ii) swaps of sensors in fog nodes. The procedure restarts from the first structure whenever the current solution is improved and continues until no improved solution can be achieved.

# 4 Experimental Results

We discuss next the performance of the VNS by evaluating it on a realistic scenario of fog computing. We start this section with the description of the experimental setup, and we proceed with the comparison between the performance of the proposed VNS and other alternatives.

## 4.1 Experimental Scenario

As a realistic fog computing scenario, we refer to a smart city project developed into the medium-sized Italian city of Modena (around 180.000 inhabitants). We consider a smart city application for monitoring car and pedestrian traffic where geographically distributed sensors collect information comprising data from proximity sensors and possibly low-resolution images. In our scenario, sensors are wireless devices located in the city's main streets: the location of the sensors is obtained by geo-referencing the selected streets. The sensors send the collected data to the fog nodes, which in turn perform pre-processing tasks by filtering the proximity sensor readings and, if available, analyze images from the camera to detect cars and pedestrians. We assume the fog nodes to be located in government buildings. The pre-processed data are then sent to a cloud data center located on the municipality premises.

In the performed experiments, we consider sensors equipped with long-range wireless connectivity, for example, LoRaWAN[1] or IEEE 802.11ah/802.11af [11]. Hence, the sensors can potentially connect to every fog node: we assume that the network delay depends on the physical distance between two nodes as in [3,4], due to the growing delay and decreasing bandwidth limitations as the distance from a sensor to the fog node increases. Specifically, we model the communication latency through the Haversine distance, starting from a given latitude and two locations' longitude.

In the experimental evaluation, we consider scenarios of different sizes to understand the proposed method's capability to scale with growing numbers of sensors and fog nodes. Specifically, we consider a number of sensors $|S| \in \{50, 100, 200\}$, and a number of fog nodes $|\mathcal{F}| \in \{5, 10, 20\}$.

We consider different scenarios, each of them described by three main parameters. First, the *sensor data rate* $\lambda$. Based on a preliminary evaluation of smart

---

[1] https://lora-alliance.org/.

city applications for traffic monitoring, we consider that each sensor provides a reading every 10 s; hence, the data rate $\lambda_i = 0.1$, $\forall i \in \mathcal{S}$. Second, the *average utilization* of the system $\rho$, that can be defined as $\frac{\sum_{i \in \mathcal{S}} \lambda_i}{\sum_{j \in \mathcal{F}} \mu_j}$. For this parameter, we consider a wide range of values: $\rho \in \{0.1, 0.2, 0.5, 0.8, 0.9\}$. For each value of $\rho$, considering sensors and fog nodes homogeneous and knowing the value of $\lambda_i$, we derive the value of $\mu_j = \mu$, which is assumed the same for each $j \in \mathcal{F}$. Third, the parameter $\delta\mu$, defining the *CPU-bound or network-bound nature* of the scenario and expressed as the ratio between the average network delay $\delta$ and the average service time of a request $(1/\mu)$. For this parameter we consider values ranging multiple orders of magnitude: $\delta\mu \in \{0.01, 0.1, 1, 10\}$. In this way, we can explore both CPU-bound scenarios (e.g., when $\delta\mu = 0.01$), where computing time is much higher than transmission time, and network-bound cases (e.g., when $\delta\mu = 10$). We derive the average network delay from the $\delta\mu$ parameter and the previously computed parameter $\mu_j$. It is worth noticing that, even if our analysis may consider very high network delays, these scenarios can still be considered realistic if we consider that the network contribution may involve the transfer of images over low-bandwidth links.

The evaluation of the proposed model considers a wide range of different scenarios related to the introduced parameters. Each scenario is named according to the format *ins-$\rho$-$\delta\mu$* (e.g., instance ins-0.1-0.01 indicates the scenario with $\rho = 0.1$ and $\delta\mu = 0.01$). Moreover, for the SLA in Eq. (15), the constant $K$ is set to 10, which is a typical value in the literature [1]. Finally, we assume the cost $c_j$ of a fog node at position $j$ equal to 1, for all $j \in \mathcal{F}$. This means that the fog nodes are equal from the operating cost point of view, and the objective function will try to reduce the overall number of active nodes. For the experimental comparison, we evaluate the following three models:

- *Simplified model (SM)*: this is the simplified version of the problem described in Sect. 2 and presented for the first time in [3]. In this model, all fog nodes are assumed on, that is $E_j = 1, \forall j \in \mathcal{F}$. The energy consumption may be high, but the infrastructure provides good performance from a response time point of view (Eq. (7));
- *Complex model (CM)*: this is the problem described in Sect. 2 that aims to minimize both energy consumption in Eq. (6) and response time in Eq. (7);
- *Proposed model (VNS)*: this is the heuristic introduced in this study and described in Sect. 3;

For the Simplified and Complex models' numeric results, we rely on Local-Solver[2] version 9.0, with a time limit of 300 s (5 min) as a stopping criterion. LocalSolver is a general mathematical programming solver that hybridizes local and direct search, constraint propagation and inference, linear and mixed-integer programming, and nonlinear programming methods. It can handle multi-objective problems, where the objectives are optimized in the order of their declaration in the model. For the sake of fairness, we run the proposed VNS heuristic for 300 s or 3000 iterations (the first to reach stops the VNS).

---

[2] http://www.localsolver.com.

**Table 2.** Results for 100 sensors and 10 available fog nodes.

| Instances | Simplified Model | | | Complex Model (Dev. CM vs. SM) | | | | | VNS (Dev. VNS vs. CM) | | | | |
|---|---|---|---|---|---|---|---|---|---|---|---|---|---|
| | Iter. | $Obj_1$ | $Obj_2$ | Iter. | $Obj_1$ | Dev. (%) | $Obj_2$ | Dev. (%) | Iter. | $Obj_1$ | Dev. (%) | $Obj_2$ | Dev. (%) |
| ins-0.1-0.01 | 23655 | 10 | 0,1163 | 52421 | 2 | -80,00 | 0,2337 | 100,96 | 1 | 2 | 0,00 | 0,2332 | -0,23 |
| ins-0.1-0.1 | 31809 | 10 | 0,1544 | 50876 | 2 | -80,00 | 0,5520 | 257,45 | 1 | 2 | 0,00 | 0,5305 | -3,90 |
| ins-0.1-1 | 29173 | 10 | 0,5219 | 61189 | 2 | -80,00 | 3,7795 | 624,22 | 1 | 2 | 0,00 | 3,2555 | -13,86 |
| ins-0.1-10 | 36088 | 10 | 4,1912 | 31853 | 6 | -40,00 | 8,6976 | 107,52 | 1 | 3 | -50,00 | 17,9568 | 106,46 |
| ins-0.2-0.01 | 26833 | 10 | 0,2613 | 25242 | 3 | -70,00 | 0,6482 | 148,07 | 1 | 3 | 0,00 | 0,6443 | -0,61 |
| ins-0.2-0.1 | 19049 | 10 | 0,3429 | 30661 | 3 | -70,00 | 1,0125 | 195,30 | 6 | 3 | 0,00 | 1,0125 | 0,00 |
| ins-0.2-1 | 28671 | 10 | 1,0829 | 33141 | 3 | -70,00 | 4,9492 | 357,05 | 4 | 3 | 0,00 | 4,5140 | -8,79 |
| ins-0.2-10 | 38641 | 10 | 8,4215 | 46185 | 3 | -70,00 | 45,6711 | 442,31 | 1 | 3 | 0,00 | 38,9263 | -14,77 |
| ins-0.5-0.01 | 39481 | 10 | 1,0300 | 13903 | 6 | -40,00 | 3,1153 | 202,46 | 1 | 6 | 0,00 | 3,1148 | -0,01 |
| ins-0.5-0.1 | 24610 | 10 | 1,2825 | 15566 | 6 | -40,00 | 3,5829 | 179,37 | 176 | 6 | 0,00 | 3,5344 | -1,35 |
| ins-0.5-1 | 21598 | 10 | 3,3132 | 7802 | 7 | -30,00 | 5,9867 | 80,70 | 86 | 6 | -14,29 | 8,1437 | 36,03 |
| ins-0.5-10 | 25093 | 10 | 21,9581 | 10851 | 7 | -30,00 | 34,4636 | 56,95 | 315 | 6 | -14,29 | 44,9171 | 30,33 |
| ins-0.8-0.01 | 52087 | 10 | 4,0480 | 11032 | 9 | -10,00 | 8,3199 | 105,53 | 40 | 9 | 0,00 | 8,3160 | -0,05 |
| ins-0.8-0.1 | 51989 | 10 | 4,4799 | 14790 | 9 | -10,00 | 8,8266 | 97,03 | 295 | 9 | 0,00 | 8,7628 | -0,72 |
| ins-0.8-1 | 38901 | 10 | 8,7654 | 14729 | 9 | -10,00 | 13,1785 | 50,35 | 305 | 9 | 0,00 | 13,2132 | 0,26 |
| ins-0.8-10 | 32297 | 10 | 44,1912 | 7335 | 9 | -10,00 | 60,2917 | 36,43 | 455 | 9 | 0,00 | 63,1833 | 4,80 |
| ins-0.9-0.01 | 57507 | 10 | 9,0540 | 11832 | 10 | 0,00 | 9,0540 | 0,00 | 16 | 10 | 0,00 | 9,0540 | 0,00 |
| ins-0.9-0.1 | 45581 | 10 | 9,5399 | 15801 | 10 | 0,00 | 9,5399 | 0,00 | 20 | 10 | 0,00 | 9,5399 | 0,00 |
| ins-0.9-1 | 54009 | 10 | 14,3987 | 10055 | 10 | 0,00 | 14,3987 | 0,00 | 16 | 10 | 0,00 | 14,3987 | 0,00 |
| ins-0.9-10 | 50609 | 10 | 62,9869 | 12502 | 10 | 0,00 | 62,9869 | 0,00 | 50 | 10 | 0,00 | 62,9869 | 0,00 |

## 4.2 Performance Evaluation

The comparison between the models performance considers two main metrics: the cost related to the number of turned on fog nodes, namely $Obj_1$, corresponding to Eq. (6); and, the average response time, $Obj_2$, corresponding to Eq. (7). To facilitate the comparison between models, we also consider a deviation measure expressing the performance of a model with respect to an alternative one. Specifically, the deviation function of a model $M1$ with respect to a model $M2$ is considered for each objective function ($Obj_1$ and $Obj_2$), which is defined as:

$$\epsilon(Obj_1^{M1}) = \frac{Obj_1^{M1} - Obj_1^{M2}}{Obj_1^{M2}} \tag{16}$$

$$\epsilon(Obj_2^{M1}) = \frac{Obj_2^{M1} - Obj_2^{M2}}{Obj_2^{M2}} \tag{17}$$

In Table 2, we present the complete results for the scenario with 100 sensors and 10 fog nodes since the others follow the same tendency. The table also shows the number of iterations required by LocalSolver and VNS to reach the reported values. Despite that, we focus the analysis on the deviation measure previously introduced to compare the models. Specifically, we consider the CM model's deviation to the SM and the deviation of the VNS to the CM.

In comparing the CM model with the SM, we note that the differences strongly depend on $\rho$. On the one hand, the SM model uses all the available fog nodes, even if, especially when $\rho$ is low, the processing of sensors data may require just a fraction of the computational infrastructure power. On the other hand, the system load has a significant impact on the number of fog nodes used

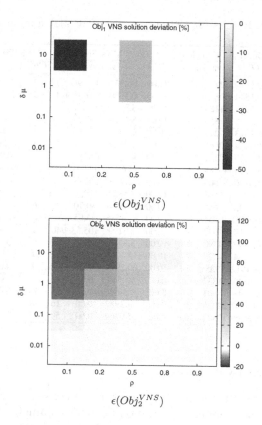

**Fig. 2.** Deviation between the VNS and CM model. (Color figure online)

by the CM model. For low values of $\rho$, indeed, the deviation on $Obj_1$ shows a reduction of the costs related to the activated fog nodes up to 80%. A higher number of active fog nodes can provide lower response time, as testified by the CM's positive deviation versus the SM on $Obj_2$. While in the SM model, we have an abundance of computational power due to all fog nodes' use, the CM uses the minimum amount of resources to satisfy the SLA constraint to reduce costs effectively.

Figures 2a and 2b have the deviation of the VNS with respect to the CM model in terms of cost ($\epsilon(Obj_1^{VNS})$) and response time ($\epsilon(Obj_2^{VNS})$), respectively. Data are represented as heat maps, with red hues when the solution performs worse than the CM model, white color when the performances are similar, and blue hues in the opposite case.

Focusing on $\epsilon(Obj_1^{VNS})$ in Fig. 2a, we observe how the VNS achieves equal or better performance for the CM model in every considered scenario. In all white areas of the chart, the number of nodes used by the VNS is the same concerning the CM model (see third and fourth columns of Table 2). Moreover, we observe that in some Network-bound scenarios (e.g., for $\rho = 0.1$ and $\delta\mu = 1$ as well as $\rho = 0.5$ and $\delta\mu \in \{1, 10\}$) the VNS allows a further reduction of fog nodes concerning the CM model, which decreases the costs up to the 50% in case of low system load ($\rho = 0.1$). The results of $\epsilon(Obj_2^{VNS})$ in Fig. 2b show that the VNS can reduce the response times in scenarios where the number of turned on fog nodes is the same as the CM model (blue areas in Fig. 2b corresponding to white areas in Fig. 2a) thanks to a more optimized mapping between sensors data flow and fog nodes. This can be explained by the fact that the CM, in the time limit of 300 s, could not reach such an optimized mapping as the proposed VNS. On the other hand, we observe some red areas corresponding to the scenarios where the number of fog nodes activated by the VNS is lower concerning the CM model: in this case, as expected, the response time increased, but it remains within the defined SLA. The effect is particularly evident for the scenario with $\rho = 0.1$ and $\delta\mu = 1$, where the VNS presents an increase in response times. Still, it can achieve a very significant reduction (50%) of the required fog nodes.

We also present two sensitivity evaluations of the proposed VNS. First, we consider a varying number of sensors $|\mathcal{S}|$ while keeping constant the number of fog nodes $|\mathcal{F}| = 100$ in Fig. 3a. Second, we evaluate how the performance changes for different sizes of scenarios, keeping constant the ratio between sensors and fog nodes ($|\mathcal{S}| = 10 \cdot |\mathcal{F}|$) and varying the number of sensors ($|\mathcal{S}| \in \{50, 100, 200\}$) in Fig. 3b.

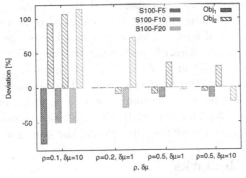

Sensitivity to the number of fog nodes $|\mathcal{F}|$.

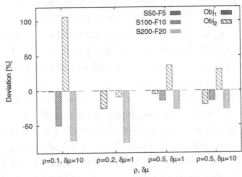

Sensitivity to number of sensors $|\mathcal{S}|$ and fog nodes $|\mathcal{F}|$.

**Fig. 3.** Sensitivity analysis of the VNS.

The results of the first sensitivity analysis are presented in Fig. 3a. They show the deviation of the VNS concerning the CM model in terms of costs ($Obj_1$) and response time ($Obj_2$) for different numbers of fog nodes, where $|\mathcal{F}|$ varies from 5 to 10 up to 20, and the number of sensors $|\mathcal{S}|$ is fixed to 100. For each case, we evaluate scenarios defined by different values of $\rho$ and $\delta\mu$. These parameters' values have been chosen among the more interesting points that emerged in the previous analysis. These are the points where the VNS behaves differently from the CM model. We observe that for a network-bound scenario with low system load ($\rho = 0.1$ and $\delta\mu = 10$), the VNS can significantly reduce the costs due to turn on fog nodes for every number of sensors, with an increase of the response times that remain within the SLA. As $\rho$ increases, we note a different behavior depending on the number of fog nodes available. If this number is low ($|\mathcal{F}| = 5$), the VNS activates all the fog nodes as the CM model; however, it achieves in some cases ($\rho = 0.5$) a reduction of the response time thanks to better mapping of sensor data flows on the fog nodes. For a higher number of fog nodes ($|\mathcal{F}| = 10, 20$), the VNS can reduce the number of required fog nodes with a low increase in the average response time.

The second sensitivity analysis is presented in Fig. 3b, where we consider systems with the ratio between the number of sensors and fog nodes fixed to 10. This analysis confirms the conclusions of the previous one, which is our proposal's stability concerning different scenarios. Also, in this case, the most significant gain in terms of required fog nodes is achieved for low system loads ($\rho \in \{0.1, 0.2\}$) and bigger sizes of the infrastructure ($|\mathcal{F}| \in \{10, 20\}$). In the smallest scenario, ($|\mathcal{S}| = 50$ and $|\mathcal{F}| = 5$), it is interesting to note that, in three scenarios out of four, the VNS can reduce the response time while using the same number of fog nodes concerning the CM model. Furthermore, in the largest scenario, ($|\mathcal{S}| = 200$ and $|\mathcal{F}| = 20$), the VNS can significantly reduce the number of required fog nodes without increasing the response time.

## 5   Concluding Remarks

The facility location problem related to the management of a fog infrastructure is investigated in this work. Specifically, we propose a VNS to solve the optimization problem of mapping sensors data flows to the fog nodes to reduce costs and response time. A qualifying point of our proposal is that starting from a list of *potential* fog nodes, it selects a minimal subset of them to guarantee the satisfaction of a service level agreement.

We test the proposed heuristic against alternative models from the literature. The VNS is evaluated in a realistic scenario of a smart city application over a wide range of scenarios characterized by different load intensities and varied nature of the application (network-bound vs. CPU-bound). The experiments show that the VNS outperforms the best alternative model in 13 out of 20 instances, finding equivalent or very close solutions in the other seven. Moreover, we perform two sensitivity analyses concerning infrastructure size.

From the sensitivity analysis, we conclude that the VNS has a very stable behavior for varying the size of the considered fog infrastructure and smart city application characteristics. In each scenario, the proposed VNS had equal or better performance concerning a state-of-the-art solver in optimizing the mapping of sensor data flows and fog nodes. It reduced the costs due to the turned-on fog nodes keeping the response time within the SLA limits.

In future works, we plan to extend the VNS to handle the fog nodes' heterogeneity and dynamic scenarios in which the load can change through time.

**Acknowledgments.** This research was partially funded by the National Counsel of Technological and Scientific Development (CNPq - grant 308312/2016-3) and the State of Goiás Research Foundation (FAPEG).

# References

1. Ardagna, D., Ciavotta, M., Lancellotti, R., Guerriero, M.: A hierarchical receding horizon algorithm for QoS-driven control of multi-IaaS applications. IEEE Trans. Cloud Comput. 1 (2018). https://doi.org/10.1109/TCC.2018.2875443
2. Canali, C., Lancellotti, R.: Scalable and automatic virtual machines placement based on behavioral similarities. Computing **99**(6), 575–595 (2016). https://doi.org/10.1007/s00607-016-0498-5
3. Canali, C., Lancellotti, R.: A fog computing service placement for smart cities based on genetic algorithms. In: Proceedings of the 9th International Conference on Cloud Computing and Services Science (CLOSER 2019), Heraklion, Greece, pp. 81–89, May 2019. https://doi.org/10.5220/0007699400810089
4. Canali, C., Lancellotti, R.: GASP: genetic algorithms for service placement in fog computing systems. Algorithms **12**(10) (2019)
5. Celik Turkoglu, D., Erol Genevois, M.: A comparative survey of service facility location problems. Ann. Oper. Res. **292**(1), 399–468 (2019). https://doi.org/10.1007/s10479-019-03385-x
6. Deng, R., Lu, R., Lai, C., Luan, T.H., Liang, H.: Optimal workload allocation in fog-cloud computing toward balanced delay and power consumption. IEEE Internet Things J. **3**(6), 1171–1181 (2016)
7. Eiselt, H.A., Laporte, G.: Objectives in location problems. In: Drezner, Z. (ed.) Facility Location: A Survey of Application and Methods, pp. 151–180. Springer, Heidelberg (1995). ISBN 978-0-387-94545-3
8. Farahani, R.Z., Fallah, S., Ruiz, R., Hosseini, S., Asgari, N.: OR models in urban service facility location: a critical review of applications and future developments. Eur. J. Oper. Res. **276**(1), 1–27 (2019)
9. Hansen, P., Mladenović, N., Todosijević, R., Hanafi, S.: Variable neighborhood search: basics and variants. EURO J. Comput. Optim. **5**(3), 423–454 (2016). https://doi.org/10.1007/s13675-016-0075-x
10. Hansen, P., Mladenović, N., Moreno Pérez, J.A.: Variable neighbourhood search: methods and applications. Ann. Oper. Res. **175**(1), 367–407 (2010)
11. Khorov, E., Lyakhov, A., Krotov, A., Guschin, A.: A survey on IEEE 802.11 ah: an enabling networking technology for smart cities. Comput. Commun. **58**, 53–69 (2015)
12. Marotta, A., Avallone, S.: A simulated annealing based approach for power efficient virtual machines consolidation. In: Proceedings of IEEE 8th International Conference on Cloud Computing (CLOUD), New York, pp. 445–452 (2015). https://doi.org/10.1109/CLOUD.2015.66

13. Mladenović, N., Alkandari, A., Pei, J., Todosijević, R., Pardalos, P.M.: Less is more approach: basic variable neighborhood search for the obnoxious p-median problem. Int. Trans. Oper. Res. **27**(1), 480–493 (2020)
14. Mladenović, N., Hansen, P.: Variable neighborhood search. Comput. Oper. Res. **24**(11), 1097–1100 (1997)
15. Sifaleras, A., Konstantaras, I.: A survey on variable neighborhood search methods for supply network inventory. In: Bychkov, I., Kalyagin, V.A., Pardalos, P.M., Prokopyev, O. (eds.) NET 2018. SPMS, vol. 315, pp. 71–82. Springer, Cham (2020). https://doi.org/10.1007/978-3-030-37157-9_5
16. Tang, B., Chen, Z., Hefferman, G., Wei, T., He, H., Yang, Q.: A hierarchical distributed fog computing architecture for big data analysis in smart cities. In: Proceedings of the ASE BigData & SocialInformatics 2015. Association for Computing Machinery, New York (2015)
17. Wen, Z., Yang, R., Garraghan, P., Lin, T., Xu, J., Rovatsos, M.: Fog orchestration for internet of things services. IEEE Internet Comput. **21**(2), 16–24 (2017)
18. Yi, S., Li, C., Li, Q.: A survey of fog computing: concepts, applications and issues. In: Proceedings of 2015 Workshop on Mobile Big Data, pp. 37–42 (2015)
19. Yousefpour, A., Ishigaki, G., Jue, J.P.: Fog computing: towards minimizing delay in the internet of things. In: 2017 IEEE International Conference on Edge Computing (EDGE), pp. 17–24 (2017)
20. Yu, V.F., Maghfiroh, M.F.: A variable neighborhood search with path-relinking for the capacitated location routing problem. J. Ind. Prod. Eng. **31**(3), 163–176 (2014)

# A GRASP/VND Heuristic
# for the Generalized Steiner Problem
# with Node-Connectivity Constraints
# and Hostile Reliability

Sebastián Laborde, Franco Robledo, Pablo Romero$^{(\boxtimes)}$, and Omar Viera

Instituto de Computación, INCO, Facultad de Ingeniería - Universidad de la
República, Montevideo, Uruguay
{nlaborde,frobledo,promero,viera}@fing.edu.uy

**Abstract.** The object under study is a combinatorial optimization problem motivated by the topological network design of communication systems, meeting reliability constraints. Specifically, we introduce the Generalized Steiner Problem with Node-Connectivity Constraints and Hostile Reliability, or GSPNCHR for short. Since the GSPNCHR belongs to the class of $\mathcal{NP}$-Hard problems, approximative algorithms are adequate for medium and large-sized networks. As a consequence, we develop a GRASP/VND methodology. The VND includes three local searches, that replace special elementary paths or trees, preserving feasibility. Our goal is to find a minimum-cost solution, meeting a reliability threshold, where both nodes and links may fail with given probabilities. We adapted TSPLIB benchmark in order to highlight the effectiveness of our proposal. The results suggest that our heuristic is cost-effective, providing highly-reliable networks.

**Keywords:** Combinatorial optimization problem · Computational complexity · Network reliability · GSPNCHR · GRASP · VND

## 1  Motivation

Currently, the backbone of the Internet infrastructure is supported by fiber-optics communication. Fiber-To-The-Home (FTTH) services have a large penetration throughout the world, and provides high data rates to the final customers. However, there are several shortcomings that should be addressed urgently. The physical design of FTTH is not suitable for large-scale natural disasters and/or malicious attacks [16]. The monitoring and detection of failures are sometimes slow, and a service disruption of hours is extremely harmful for business models. Furthermore, FTTH services are suffering the capacity crunch problem, and elastic optical networks combined with smart traffic engineering and additional redundancy is currently in order.

© Springer Nature Switzerland AG 2021
N. Mladenovic et al. (Eds.): ICVNS 2021, LNCS 12559, pp. 43–57, 2021.
https://doi.org/10.1007/978-3-030-69625-2_4

Consequently, a smart augmentation of the physical network is mandatory. Since the deployment of fiber-optics is an important economical investment, the topological network design of FTTH networks should be revisited. The goal is to interconnect distinguished nodes, called terminals, using large level of redundancy, and simultaneously, meeting large reliable constraints.

Reliability analysis deals with probabilistic failures on the components of a system. The reliability is precisely the probability of correct operation of the whole system, subject to random failures. Here, we consider a realistic *hostile model*, where both nodes and links could fail. Our goal is to understand the cost-reliability trade-off, and how the reliability is naturally increased adding levels of redundancy between distinguished terminals. The contributions of this paper can be summarized in the following items:

1. The Generalized Steiner Problem with Node-Connectivity Constraints and Hostile Reliability (GSPNCHR), is introduced.
2. We formally prove that the GSPNCHR belongs to the $\mathcal{NP}$-Hard class.
3. As a consequence, a GRASP/VND methodology is proposed.
4. Our results highlight that the model is robust under non-terminal node-failures, rather than link-failures.

The document is organized in the following manner. The related work is presented in Section 2. A formal description for the GSPNCHR is presented in Section 3; its $\mathcal{NP}$-Hardness is also established. A GRASP/VND solution is introduced in Section 4. Numerical results are presented in Section 5. Section 6 contains concluding remarks and trends for future work.

## 2   Related Work

Here, we extend the Generalized Steiner Problem (GSP), adding node-connectivity requirements and a hostile network reliability model with probabilistic failures on its components. We cover the main body of related works in the fields of network reliability analysis, topological network design and joint problems from the scientific literature.

Scarce works jointly deal with a topological network optimization under reliability constraints. J. Barrera et al. proposed a topological network optimization, trying to minimize costs subject to $K$-terminal reliability constraints [3]. The authors consider Sample Average Approximation method, which is a powerful tool for stochastic optimization and for the resolution of $\mathcal{NP}$-Hard combinatorial problems with a target probability maximization [13]. They conclude that suboptimal solutions could be found if dependent failures are ignored in the model. The scientific literature also offers topological optimization problems meeting reliability constraints, or reliability maximization under budget constraints, which is known as network synthesis. The reader can find a survey on the synthesis in network reliability in [4]. More recent works propose a reliability optimization in general stochastic binary systems [7], even under the introduction of Sample Average Approximation [20]. Building uniformly most-reliable graphs is an active and challenging research field, where the goal is to find graphs with fixed nodes and links with maximum reliability evaluation in a uniform sense. The interested reader can consult [2] for conjectures in this field. A close problem to ours is to consider topological transformations (i.e., moving links or path/tree replacements) in order to increase the reliability measure. This problem is not mature, and a recent work propose a novel reliability-increasing network transformation [6]. There, E. Canale et al. show that any graph with a cut-point can be transformed into a biconnected graph with greater reliability. Using this remarkable property, our design does not include cut-points.

Most works in the field of network reliability analysis deal with its evaluation rather than its maximization. The literature on network reliability evaluation is abundant, and here we can mention distinguished works on this field. A trade-off between accuracy and computational feasibility is met by simulations, given the hardness of the classical network reliability models [11]. Macroscopically, Monte Carlo methods consider independent replications of a complex system, and by means of statistical laws find pointwise estimations, in order to make decisions on the system. The reader is invited to consult an excellent book on Monte Carlo methods authored by Fishman [9], which was inspirational for network reliability, numerical integration, statistics and other fields of knowledge. In our particular case we deal with the hostile network reliability model, where both links and non-terminal nodes fail independently. Its reliability evaluation belongs to the class of $\mathcal{NP}$-Hard problems as well [11]. Recursive Variance Reduction (RVR) is an outstanding technique for the reliability estimation [8]. This formulation allows a meaningful variance reduction, and the product between time and variance is also reduced when compared to Crude Monte Carlo (CMC). Furthermore, the variance is mathematically proved to be always better in RVR than in CMC. More recently, the applicability of RVR is extended to Stochastic Monotone Binary Systems (SMBS) [5]. Since the hostile model is a SMBS, in this work we consider RVR for the network reliability estimation.

In the Generalized Steiner Problem (GSP), the goal is to communicate a given subset of terminal-nodes at the minimum cost, meeting connectivity requirements either by link-disjoint (GSP-EC) or node-disjoint (GSP-NC) paths. Since the problem is $\mathcal{NP}$-Hard, the literature offers approximation algorithms as well as metaheuristics. K. Agrawal and Ravi [1] developed an approximation algorithm with logarithmic factor for the GSP-EC. Jain [12] presented a factor-2 approximation algorithm for the GSP-EC, using the primal-dual schema for linear programming. A deep inapproximability result for the GSP-NC without Steiner nodes was introduced by Kortsarz in [14]. In [24] an enumeration of optimal solutions for the GSP is carried out with a compact data structure, called Zero-Suppressed Binary Decision Diagrams. The authors show that this method works for several real-world instances. Heuristics are also available for the GSP. Sartor and Robledo developed a GRASP methodology to address the GSP-EC [23]. In [19], S. Nesmachnow presents an empirical evaluation of several simple metaheuristics (VNS is included) to address the GSP, with promising results. Several implementations of VNS have been developed for the particular

Traveling Salesman Problem as well, showing that VNS is competitive. Exact solutions for the TSP and extensions can also be found in [22]. The CPU-times provided by the exact solutions are longer than the heuristics, and in particular VNS proposal for the TSP.

# 3   Problem and Complexity

In this section, we first present a general description of the GSPNCHR. Then, a formal combinatorial optimization problem is introduced, and the hardness is established.

**Definition 1 (GSPNCHR).** *Consider a simple undirected graph $G = (V, E)$, terminal-set $T \subseteq V$, link-costs $\{c_{i,j}\}_{(i,j) \in E}$ and connectivity requirements $R = \{r_{i,j}\}_{i,j \in T}$. Further, we assume that both links and non-terminal (Steiner) nodes fail with respective probabilities $P_E = \{p_e\}_{e \in E}$ and $P_{V-T} = \{p_v\}_{v \in V-T}$. Given a reliability threshold $p_{min}$, the goal is to build a minimum-cost topology $G_S \subseteq G$ meeting both the connectivity requirements $R$ and the reliability threshold: $R_K(G_S) \geq p_{min}$, being $K = T$ the terminal-set.*

Recall that the $K$-Terminal reliability is the probability that all the terminals belong to the same component, after node and link failures. The exact computation of the reliability $R_K(G)$ is $\mathcal{NP}$-Hard [11]. Consider an instance $(G, C, R, T, P_E, P_{V-T}, p_{min})$ of the GSPNCHR, and the following decision variables:

$$y_{(i,j)}^{u,v} = \begin{cases} 1 \ if(i,j) \in E \ is \ used \ in \ a \ path \ u - i - j - v \\ 0 \ otherwise \end{cases}$$

$$x_{(i,j)} = \begin{cases} 1 \ if(i,j) \in E \ is \ used \ in \ the \ solution \\ 0 \ otherwise \end{cases}$$

$$\hat{x}_i = \begin{cases} 1 \ if \ the \ Steiner \ node \ i \in V - T \ is \ used \ in \ the \ solution \\ 0 \ otherwise \end{cases}$$

Here, we introduce the GSPNCHR as the following combinatorial optimization problem:

$$\min \sum_{(i,j) \in E} c_{i,j} x_{i,j}$$

$$s.t. \ x_{ij} \geq y_{(i,j)}^{u,v} + y_{(j,i)}^{u,v} \ \forall (i,j) \in E, \forall u, v \in T, u \neq v \quad (1)$$

$$\sum_{(u,i) \in E} y_{(u,i)}^{u,v} \geq r_{u,v} \ \forall u, v \in T, \ u \neq v \quad (2)$$

$$\sum_{(j,v) \in E} y_{(j,v)}^{u,v} \geq r_{u,v} \ \forall u, v \in T, \ u \neq v \quad (3)$$

$$\sum_{(i,p) \in I(p)} y_{(i,p)}^{u,v} - \sum_{(p,j) \in I(p)} y_{(p,j)}^{u,v} \geq 0, \ \forall p \in V - \{u, v\}, \forall u, v \in T, u \neq v \quad (4)$$

$$\sum_{(s,i)\in E} x_{s,i} \leq M\hat{x}_s, \forall s \in V - T \tag{5}$$

$$R_K(G_S) \geq p_{min} \tag{6}$$

$$x_{(i,j)} \in \{0,1\} \forall (i,j) \in E \tag{7}$$

$$\hat{x}_i \in \{0,1\} \forall i \in V - T \tag{8}$$

$$y_{(i,j)}^{u,v} \in \{0,1\} \forall (i,j) \in E, \forall u,v \in T, u \neq v \tag{9}$$

The goal is to minimize the global cost of the solution. The set of Constraints 1 state that links are one-way. The connectivity requirements are expressed by means of Constraints 2 and 3. Constraints 4 represent Kirchhoff law, or flow conservation. Constraints 5 state that an incident link to a Steiner node can be used only if the Steiner node is considered in the solution. Observe that $M$ is a large real number; $M = |E|$ can be used in the model without loss of generality. The minimum reliability threshold is established with Constraint 6, being $G_S \subseteq G$ the subgraph with all the constructed links $x_{i,j}$. Finally, the set of constraints 7-9 state that all the decision variables belong to the binary set $\{0,1\}$. Now, we establish the hardness for the GSPNCHR.

**Theorem 1.** *The GSPNCHR belongs to the class of $\mathcal{NP}$-Hard problems.*

*Proof.* By inclusion. Recall that Hamilton Tour belongs to Karp list of $\mathcal{NP}$-Complete problems [11]. Consider a simple graph $G = (V, E)$. Consider the *trivial instance* $(G, C, R, T, P_E, P_{V-T}, p_{min})$ with unit costs, perfect nodes/links, no Steiner nodes and requirements $r_{i,j} = 2$ for all $i, j \in V$. The cost is not greater than $n = |V|$ if and only if $G$ has a Hamilton tour.

The GSPECHR is also $\mathcal{NP}$-Hard; the proof is analogous. Theorem 1 can be strengthened considering strong inapproximability results [10].

In order to tackle the GSPNCHR, first we provide full solution for the relaxed version without the reliability threshold, this is, without Constraint 6. Then, we count the number of feasible solutions that meet that constraint. We use this approach, since we want to determine if the topological robustness has an impact in the resulting network reliability.

## 4    Algorithmic Proposal

GRASP and VND are well known metaheuristics that have been successfully used to solve many hard combinatorial optimization problems. GRASP is a powerful multi-start process which operates in two phases [21]. A feasible solution is built in a first phase, whose neighborhood is then explored in the Local Search Phase [21]. The second phase is usually enriched by means of different variable neighborhood structures. For instance, VND explores several neighborhood structures in a deterministic order. Its success is based on the simple fact that different neighborhood structures do not usually have the same local minimum. Thus, the resulting solution is simultaneously a locally optimum solution under all the neighborhood structures. The reader is invited to consult the comprehensive Handbook of Heuristics for further information [17]. Here, we develop a GRASP/VND methodology.

### 4.1   General Scheme

A pseudocode of our full proposal is presented in Fig. 1. It receives the ground graph $G_B$, a number of iterations $iter$ and a positive integer $k$ to find the $k$ shortest paths during the Construction Phase, a reliability threshold $p_{min}$, the elementary reliabilities $P_E$, $P_{V-T}$ and number of iterations $simiter$ during the simulations carried out in the Reliability Phase. If the resulting solution respects Constraint 6, it is included in the set $sol$, that is returned in Line 10. Observe that RVR method is considered in order to test this reliability constraint [8]. Our goal is to determine how many solutions for the relaxed problem meet Constraint 6, as a function of the robustness (connectivity matrix $R$, elementary reliabilities and threshold $p_{min}$).

---

**Algorithm 1** $sol = NetworkDesign(G_B, iter, k, p_{min}, P_E, P_{V-T}, simiter)$

1: $i \leftarrow 0$; $P \leftarrow \emptyset$; $sol \leftarrow \emptyset$
2: **while** $i < iter$ **do**
3:     $\overline{g} \leftarrow Construction(G_B, P, k)$
4:     $g_{sol} \leftarrow VND(\overline{g}, P)$
5:     $reliability \leftarrow RVR(g_{sol}, P_E, P_{V-T}, simiter)$
6:     **if** $reliability > p_{min}$ **then**
7:         $sol \leftarrow sol \cup \{g_{sol}\}$
8:     **end if**
9: **end while**
10: **return** $sol$

---

**Fig. 1.** Pseudocode for the main algorithm: $NetworkDesign$.

### 4.2   Construction Phase

This algorithm trades simplicity and effectiveness, building paths iteratively. Figure 2 receives the ground graph $G_B$, the matrix with link-costs $C$, the connectivity matrix $R$, and the parameter $k$. Denote $S_D^{(I)}$ the set of terminal nodes, following the terminology of the backbone design from Wide Area Networks. In Line 1, the solution $g_{sol}$ is initialized only with the terminal nodes $S_D^I$ without links, $M = \{m_{i,j}\}_{i,j \in T}$ stores the unsatisfied requirements, so initially $m_{i,j} = r_{i,}$ for all $i, j \in S_D^{(I)}$, and the matrix $P = \{P_{i,j}\}_{i,j \in S_D^{(I)}}$ that represents the collection of node-disjoint paths is empty for all $P_{i,j}$. Additionally, the matrix $A = \{A_{i,j}\}_{i,j \in S_D^{(I)}}$ that controls the number of attempts that the algorithm fails to find $r_{i,j}$ node-disjoint paths between $i, j$ is initialized correspondingly: $A_{i,j} = 0\ \forall i, j \in S_D^{(I)}$.

---

**Algorithm 2** $(sol, P) = Construction(G_B, C, R, k)$

1: $g_{sol} \leftarrow (S_D^{(I)}, \emptyset); \ m_{i,j} \leftarrow r_{i,j}; \ P_{i,j} \leftarrow \emptyset, \forall i,j \in S_D^{(I)}; \ A_{i,j} \leftarrow 0, \forall i,j \in S_D^{(I)}$
2: **while** $\exists m_{i,j} > 0 : A_{i,j} < MAXATTEMPTS$ **do**
3:      $(i,j) \leftarrow ChooseRandom(S_D^{(I)} : m_{i,j} > 0)$
4:      $\overline{G} \leftarrow G_B \setminus P_{i,j}$
5:      **for all** $(u,v) \in E(\overline{G})$ **do**
6:          $\overline{c}_{u,v} \leftarrow c_{u,v} \times 1_{\{(u,v)\notin g_{sol}\}}$
7:      **end for**
8:      $L_p \leftarrow KSP(k, i, j, \overline{G}, \overline{C})$
9:      **if** $L_p = \emptyset$ **then**
10:        $A_{i,j} \leftarrow A_{i,j} + 1; \ P_{i,j} \leftarrow \emptyset; \ m_{i,j} \leftarrow r_{i,j}$
11:      **else**
12:        $p \leftarrow SelectRandom(L_p); \ g_{sol} \leftarrow g_{sol} \cup \{p\}$
13:        $P_{i,j} \leftarrow P_{i,j} \cup \{p\}; \ m_{i,j} \leftarrow m_{i,j} - 1$
14:        $(P, M) \leftarrow GeneralUpdateMatrix(g_{sol}, P, M, p, i, j)$
15:      **end if**
16: **end while**
17: **return** $(g_{sol}, P)$

---

**Fig. 2.** Pseudocode for the Construction Phase.

The purpose of the *while-loop* (Lines 2-13) is to fulfill all the connectivity requirements in a randomized fashion. Observe that we selected a large Restricted Candidate List (RCL) in our GRASP proposal for diversification purposes. A pair of terminals $(i, j)$ is uniformly picked at random from the set $S_D^{(I)}$, provided that $m_{i,j} > 0$ (Line 3). The graph $\overline{G}$ defined in Line 4 discards the nodes that were already visited in the previous paths. Therefore, if we find some path between $i$ and $j$ in $\overline{G}$, it will be included. In the *for-loop* of Lines 5-7, an auxiliary matrix with the costs $\overline{C} = \overline{c}_{i,j}$ allows to use already existent links from $g_{sol}$ without additional cost, and add them to build a new node-disjoint path. The KSP from $i$ to $j$ are computed in Line 8 using Yen algorithm [18], that finds the $k$-Shortest Paths between two fixed nodes in a graph. In Line 9, we test if the list $L_p$ is empty. In this case we re-initialize $P_{i,j}$, $m_{i,j}$, and add a unit to $A_{i,j}$, since $i$ and $j$ belong to different connected components. If the list $L_p$ is not empty, a path $p$ is uniformly picked from the list $L_p$, and it is included in the solution (Line 12). The path $p$ is added to $P_{i,j}$, and the requirement $m_{i,j}$ is decreased a unit (Line 13). The addition of the path $p$ could build node-disjoint paths from different terminals. Consequently, the function *GeneralUpdateMatrix* finds these new paths. *Construction* returns a feasible solution $g_{sol}$ equipped with all the sets $P = \{P_{i,j}\}_{i,j \in S_D^{(I)}}$ of node-disjoint pairs between the different terminals (Line 17). The reader can observe that *Construction* returns a feasible solution for the GSPNC, which is the relaxed version of the GSPNCHR.

## 4.3   Local Search Phase - VND

The goal is to combine a rich diversity of neighborhoods in order to obtain an output that is locally optimum solution for every feasible neighborhood. Here, we consider three neighborhood structures to build a VND [17]. First, the concept of key-nodes, key-paths and key-trees are in order:

**Definition 2 (key-node).** *A key-node in a feasible solution $v \in g_{sol}$ is a Steiner (non-terminal) node with degree three or greater.*

**Definition 3 (key-path).** *A key-path in a feasible solution $p \subseteq g_{sol}$ is an elementary path where all the intermediate nodes are non-terminal with degree 2 in $g_{sol}$, and the extremes are either terminals or key-nodes.*

A feasible solution $g_{sol}$ accepts a decomposition into key-paths: $K_{g_{sol}} = \{p_1, \ldots, p_h\}$.

**Definition 4 (key-tree).** *Let $v \in g_{sol}$ be a key-node belonging to a feasible solution $g_{sol}$. The key-tree associated to $v$, denoted by $T_v$, is the tree composed by all the key-paths that meet in the common end-point (i.e., the key-node $v$).*

Now, we are in conditions to define three neighborhood structures that combine the previous concepts. Consider a feasible solution $g_{sol}$ for the GSPNC.

**Definition 5 (Neighborhood Structure for swap key-paths).** *Given a key-path $p \subseteq g_{sol}$, a neighbor solution for $g_{sol}$ is $\hat{g}_{sol} = \{g_{sol} \setminus p\} \cup \{m\}$, being $m$ the set of nodes and links that will be added to preserve the feasibility of $\hat{g}_{sol}$.*

**Definition 6 (Neighborhood Structure for key-paths).** *Given a key-path $p \in g_{sol}$, a neighbor-solution is $\hat{g}_{sol} = \{g_{sol} \setminus p\} \cup \{\hat{p}\}$, where $\hat{p}$ is other path that connects the extremes from $p$. The neighborhood of key-paths from $g_{sol}$ is composed by the previous operation to the possible members belonging to $K_{g_{sol}}$.*

**Definition 7 (Neighborhood Structure for key-tree).** *Consider the key-tree $T_v \in g_{sol}$ rooted at the key-node $v$. A neighbor of $g_{sol}$ is $\hat{g}_{sol} = \{g_{sol} \setminus T_v\} \cup \{T\}$, being $T$ another tree that replaces $T_v$ with identical leaf-nodes.*

Our full algorithm *NetworkDesign* considers a classical VND implementation, calling the three respective local searches in order, after the *Construction* phase:

1. *SwapKeyPathLocalSearch*
2. *KeyPathLocalSearch*
3. *KeyTreeLocalSearch*

Even though the last two local searches are the most simple, preliminary experiments suggested that *SwapKeyPathLocalSearch* should take effect first. The respective pseudocodes for the different local searches are presented in Figs. 4-3. It is worth to remark that these local searches take effect only if the resulting solution is both feasible and cheaper than the original one. The respective codes from each local search are self-explanatory, and strictly follow the corresponding neighborhood structures, trying to find better replacements. For completeness, two auxiliary functions called during these searches are here explained, in terms of inputs and outputs. We invite the reader to consult [15] for implementation details:

- $GeneralRecConnect$: receives the ground graph $G_B$, cost-matrix $C$, current solution $g_{sol}$ and a key-node $v$. It tries to replace the key-tree $T_v$ with a better key-tree $T$ spanning the same leaf-nodes, preserving feasibility. It returns another solution and a boolean $improve$ (if $improve = 0$, an identical solution is returned).
- $FindSubstituteKeyPath$: receives the current solution $g_{sol}$, the key-path $p$ and a matrix $P$ with the collection of disjoint path between the terminals. It replaces the current path $p$ by $\hat{p}$, exploiting the information given by $P$ in order to reconstruct a new feasible solution. If this solution is cheaper, it returns $improve = 1$ and the resulting solution (otherwise, an identical solution is returned).

In the last step, RVR is introduced in order to determine if Constraint 6 is met. The reader is invited to consult authoritative works on RVR and cites therein [8] (Fig. 5).

---

**Algorithm 3** $g_{sol} = SwapKeyPathLocalSearch(G_B, C, g_{sol}, P)$

1: $improve \leftarrow TRUE$
2: **while** $improve$ **do**
3:     $improve \leftarrow FALSE$
4:     $K(g_{sol}) \leftarrow \{p_1, \ldots, p_h\}$ /* Key-path decomposition of $g_{sol}$ */
5:     **while not** $improve$ **and** $\exists$ **key-paths not analyzed do**
6:         $p \leftarrow (K(g_{sol}))$ /* Path not analyzed yet */
7:         $(g_{sol}, improve) \leftarrow FindSubstituteKeyPath(g_{sol}, p, P)$
8:     **end while**
9: **end while**
10: **return** $g_{sol}$

---

**Fig. 3.** Pseudocode for Local Search 3: $SwapKeyPathLocalSearch$.

---

**Algorithm 4** $g_{sol} = KeyPathLocalSearch(G_B, C, g_{sol})$

---

1: $improve \leftarrow TRUE$
2: **while** $improve$ **do**
3:     $improve \leftarrow FALSE$
4:     $K(g_{sol}) \leftarrow \{p_1, \ldots, p_h\}$ /* Key-path decomposition of $g_{sol}$ */
5:     **while not** $improve$ **and** $\exists$ **key-paths not analyzed do**
6:         $p \leftarrow (K(g_{sol}))$ /* Path not analyzed yet, with extremes $u$ and $v$ */
7:         $\hat{\mu} \leftarrow < NODES(p) \cup S_D \setminus NODES(g_{sol}) >$ /* Induced subgraph $\hat{\mu}$ */
8:         $\hat{p} \leftarrow Dijkstra(u, v, \hat{\mu})$
9:         **if** $COST(\hat{p}) < COST(p)$ **then**
10:             $g_{sol} \leftarrow \{g_{sol} \setminus p\} \cup \{\hat{p}\}$
11:             $improve \leftarrow TRUE$
12:         **end if**
13:     **end while**
14: **end while**
15: **return** $g_{sol}$

---

**Fig. 4.** Pseudocode for Local Search 1: $KeyPathLocalSearch$.

---

**Algorithm 5** $g_{sol} = KeyTreeLocalSearch(G_B, C, g_{sol})$

---

1: $improve \leftarrow TRUE$
2: **while** $improve$ **do**
3:     $improve \leftarrow FALSE$
4:     $X \leftarrow KeyNodes(g_{sol})$ /* Key-nodes from $g_{sol}$ */
5:     $\overline{S} \leftarrow S_D \setminus NODES(g_{sol})$
6:     **while not** $improve$ **and** $\exists$ **key-nodes not analyzed do**
7:         $v \leftarrow X$ /* Key-node not analyzed yet */
8:         $[g_{sol}, improve] \leftarrow GeneralRecConnect(G_B, C, g_{sol}, v, \overline{S})$
9:     **end while**
10: **end while**
11: **return** $g_{sol}$

---

**Fig. 5.** Pseudocode for Local Search 2: $KeyTreeLocalSearch$.

## 5   Numerical Results

In order to understand the effectiveness of this proposal, an extensive computational study was carried out using our main algorithm $NetworkDesign$. The experimental analysis was carried out in a laptop (Pentium Core I5, 8GB). Since there are no benchmark for our specific problem we adapted the well-known TSPLIB instances, adding node/link failure probabilities and node connectivity requirements. We selected $k = 5$ for $Construction$, which showed acceptable results in a training set. In our reliability-centric design, we fixed $p_{min} = 0.8$; lower values make no sense. The elementary reliabilities for both Steiner nodes and links are close to the unit, since we are focused on the design of highly-reliable networks. Specifically, the nine combinations for $p_v, p_e \in \{0.99, 0.97, 0.95\}$ were

**Table 1.** GRASP/VND effectiveness

| Instance | % T | % IC | % IVND | CPU (s) | $\overline{R}$ | $\overline{Var}$ |
|---|---|---|---|---|---|---|
| att48 | 20 | 99.27 | 34.61 | 11.466 | 0.967 | 7.608E−07 |
| att48 | 35 | 98.6 | 36.83 | 29.769 | 0.943 | 3.448E−06 |
| att48 | 50 | 98.22 | 37.1 | 65.904 | 0.927 | 5.322E−06 |
| berlin52 | 20 | 98.98 | 30.55 | 30.605 | 0.937 | 3.294E−06 |
| berlin52 | 35 | 99.06 | 33.93 | 33.433 | 0.938 | 3.19E−06 |
| berlin52 | 50 | 98.02 | 33.48 | 106.945 | 0.907 | 6.487E−06 |
| brazil58 | 20 | 98.92 | 31.96 | 62.377 | 0.885 | 6.722E−06 |
| brazil58 | 35 | 99.25 | 39.45 | 68.891 | 0.86 | 8.347E−06 |
| brazil58 | 50 | 98.75 | 35.26 | 103.553 | 0.91 | 7.093E−06 |
| ch150 | 20 | 99.76 | 37.51 | 222.552 | 0.8559 | 1.029E−05 |
| ch150 | 35 | 99.72 | 36.65 | 546.652 | 0.8803 | 9.033E−05 |
| gr202 | 20 | 99.89 | 32.43 | 100.162 | 0.8231 | 1.224E−05 |
| gr202 | 35 | 99.75 | 34.56 | 200.698 | 0.8414 | 1.11E−05 |
| gr202 | 50 | 99.74 | 33.36 | 600.629 | 0.8303 | 1.279E−05 |
| rd400 | 20 | 99.94 | 35.84 | 88.214 | 0.8094 | 14.22E−05 |
| rd400 | 35 | 99.94 | 33.54 | 504.103 | 0.8537 | 11.89E−05 |
| rd400 | 50 | 99.93 | 33.16 | 980.701 | 0.8643 | 11.51E−05 |
| Average | 35 | 99.28 | 34.72 | 220.980 | 0.884 | 3.28E−05 |

considered in different instances, being $p_v$ and $p_e$ the elementary reliabilities for Steiner nodes and links $e = (i, j)$ respectively. The number of iterations for *NetworkDesign* is established in *iter* = 100, and the number of iterations for the RVR method is $10^4$. We want to understand the sensibility of the solution to perturbations in the elementary reliabilities. Therefore, different values for the elementary reliabilities for both Steiner nodes and links were used. Table 1 shows the results for each adapted TSPLIB instance. Each column contains, respectively, name of the TSPLIB instance, percentage of terminal nodes (% *T*), relative improvements of *Construction* (% *IC*) and *VND* phases (%*IVND*), in relation to the cost of the corresponding input graphs, CPU-time per iteration of *NetworkDesign*, reliability estimation $\overline{R}$ and estimated variance $\overline{Var}$. From Table 1, we can appreciate that the cost of the resulting graph after the *Construction* is practically one-half the cost of its input. The improvement of *VND* is consistently bounded between 30.55% and 39.45%, according to the instance and its characteristics on the test-set. The minimum threshold $p_{min} = 0.8$ is widely exceeded in all the instances under study, considering the RVR estimation $\overline{R}$. The elementary reliabilities were established in $p_v = 0.99$ and $p_e = 0.95$ respectively for nodes and links, and the range of reliabilities

**Table 2.** Feasible solutions with $R \geq 0.98$, $p_v = 0.99$ fixed and variable link reliability

| Instance | $p_e = 0.99$ | $p_e = 0.97$ | $p_e = 0.95$ |
|---|---|---|---|
| att48 T20 | 100 | 90 | 12 |
| att48 T35 | 100 | 53 | 0 |
| att48 T50 | 100 | 20 | 0 |
| berlin52 T20 | 100 | 41 | 0 |
| berlin52 T35 | 100 | 50 | 0 |
| berlin52 T50 | 100 | 1 | 0 |
| brazil58 T20 | 99 | 15 | 0 |
| brazil58 T35 | 97 | 0 | 0 |
| brazil58 T50 | 100 | 5 | 0 |
| ch150 T20 | 100 | 0 | 0 |
| ch150 T35 | 100 | 0 | 0 |
| ch150 T50 | 100 | 0 | 0 |
| gr202 T20 | 99 | 0 | 0 |
| gr202 T35 | 100 | 0 | 0 |
| gr202 T50 | 100 | 0 | 0 |
| rd400 T20 | 100 | 0 | 0 |
| rd400 T35 | 100 | 0 | 0 |
| rd400 T50 | 100 | 0 | 0 |
| Average | 99.72 | 15.28 | 0.67 |

is bounded between 0.8094 and 0.967, meeting the reliability constraint. The estimated variance $\overline{Var}$ is reduced in average in all the instances under study. These facts highlight the activity of the $VND$ phase, the accuracy of RVR and the global effectiveness of our proposal. Furthermore, the CPU times are acceptable, even under large-sized graphs with 400 nodes.

Tables 2 and 3 illustrate the number of feasible solutions obtained when we fix the elementary node-reliability and modify the link-reliabilities, and vice-versa. The suffix $TXY$ in each instance indicates the percentage $XY\%$ of terminal nodes. The feasibility is 100% in almost all the instances under study when both $p_e = p_v = 0.99$. However, the feasibility is dramatically deteriorated as soon as the link-reliabilities are decreased (see the last column of Table 2). This effect is not pronounced when the node-reliabilities are reduced, as we can appreciate from the last column of Table 3. This fact shows that the system is robust under failures of Steiner nodes.

**Table 3.** Feasible solutions with $R \geq 0.98$, $p_e = 0.99$ fixed and variable node-reliability

| Instance | $p_v = 0.99$ | $p_v = 0.97$ | $p_v = 0.95$ |
|---|---|---|---|
| tt48 T20 | 100 | 100 | 99 |
| att48 T35 | 100 | 98 | 96 |
| att48 T50 | 100 | 100 | 99 |
| berlin52 T20 | 100 | 100 | 80 |
| berlin52 T35 | 100 | 99 | 93 |
| berlin52 T50 | 100 | 100 | 100 |
| brazil58 T20 | 99 | 59 | 41 |
| brazil58 T35 | 97 | 43 | 9 |
| brazil58 T50 | 100 | 99 | 81 |
| ch150 T20 | 100 | 60 | 20 |
| ch150 T35 | 100 | 98 | 76 |
| ch150 T50 | 100 | 100 | 97 |
| gr202 T20 | 99 | 80 | 30 |
| gr202 T35 | 100 | 69 | 16 |
| gr202 T50 | 100 | 100 | 76 |
| rd400 T20 | 100 | 16 | 2 |
| rd400 T35 | 100 | 98 | 80 |
| rd400 T50 | 100 | 100 | 100 |
| Average | 99.72 | 84.39 | 66.39 |

# 6   Conclusions and Trends for Future Work

We studied the topological design of highly reliable networks. Our goal is to combine purely deterministic aspects such as connectivity with probabilistic models coming from network reliability. For that purpose, the Generalized Steiner Problem with Node-Connectivity Constraints and Hostile Reliability (GSPNCHR) is here introduced. The GSPNCHR belongs to the class of $\mathcal{NP}$-Hard problems, since it subsumes the Generalized Steiner Problem (GSP). Therefore, exact methods are prohibitive, even for networks with moderate size. A GRASP/VND solution is here developed, which shows to be both flexible and effective. Since the reliability evaluation for the hostile model also belongs to the $\mathcal{NP}$-Hard class, we adopted an outstanding pointwise reliability estimation, known as Recursive Variance Reduction (RVR) method. This method is unbiased, accurate and it presents small variance, as the results show. The model is more sensible to link-failures rather than node-failures.

The interplay between topological network design and network reliability is not well understood yet. Some local searches were here proposed, essentially using key-path and key-tree replacements, in order to reduce costs preserving feasibility. A current research line is to introduce reliability-increasing transformations. The development of local searches that increase reliability and reduce costs would enrich the current solution. Another possibility for future work is to enrich the number of local searches and consider probabilistic transitions between them.

**Acknowledgements.** This work is partially supported by Project ANII FCE_1_2019_ 1_156693 *Teoría y Construcción de Redes de Máxima Confiabilidad*, MATHAMSUD 19-MATH-03 *Rare events analysis in multi-component systems with dependent components* and STIC-AMSUD ACCON *Algorithms for the capacity crunch problem in optical networks*.

# References

1. Agrawal, A., Klein, P., Ravi, R.: When trees collide: an approximation algorithm for the generalized Steiner problem on networks. SIAM J. Comput. **24**(3), 440–456 (1995)
2. Archer, K., Graves, C., Milan, D.: Classes of uniformly most reliable graphs for all-terminal reliability. Discret. Appl. Math. **267**, 12–29 (2019)
3. Barrera, J., Cancela, H., Moreno, E.: Topological optimization of reliable networks under dependent failures. Oper. Res. Lett. **43**(2), 132–136 (2015)
4. Boesch, F.T., Satyanarayana, A., Suffel, C.L.: A survey of some network reliability analysis and synthesis results. Networks **54**(2), 99–107 (2009)
5. Canale, E., et al.: Recursive variance reduction method in stochastic monotone binary systems. In: Proceedings of the 7th International Workshop on Reliable Networks Design and Modeling, pp. 135–141 (2015)
6. Canale, E., Robledo, F., Romero, P., Viera, J.: Building reliability-improving network transformations. In: Proceedings of the 15th International Conference on the Design of Reliable Communication Networks, pp. 107–113. IEEE (2019)
7. Cancela, H., Guerberoff, G., Robledo, F., Romero, P.: Reliability maximization in stochastic binary systems. In: Proceedings of the 21st Conference on Innovation in Clouds, Internet and Networks and Workshops, pp. 1–7. IEEE (2018)
8. Cancela, H., El Khadiri, M., Rubino, G.: A new simulation method based on the RVR principle for the rare event network reliability problem. Ann. Oper. Res. **196**(1), 111–136 (2012)
9. Fishman, G.: Monte Carlo. Springer Series in Operations Research and Financial Engineering. Springer, Heidelberg (1996)
10. Gabow, H., Goemans, M., Williamson, D.: An efficient approximation algorithm for the survivable network design problem. Math. Program. **82**, 13–40 (1998)
11. Garey, M., Johnson, D.: Computers and Intractability; A Guide to the Theory of NP-Completeness. W. H. Freeman and Co., USA (1990)
12. Kamal, J.: A factor 2 approximation algorithm for the generalized Steiner network problem. In: Proceedings of the 39th Annual Symposium on Foundations of Computer Science, FOCS 1998, p. 448. IEEE Computer Society (1998)
13. Kleywegt, A., Shapiro, A., Homem-de-Mello, T.: The sample average approximation method for stochastic discrete optimization. SIAM J. Optim. **12**(2), 479–502 (2002)
14. Kortsarz, G., Krauthgamer, R., Lee, J.: Hardness of approximation for vertex-connectivity network design problems. SIAM J. Comput. **33**(3), 704–720 (2004)

15. Laborde, S.: Topological Optimization of Fault-Tolerant Networks meeting Reliability Constraints. Master's thesis, Universidad de la República (2020)
16. Lourenço, R., Figueiredo, G., Tornatore, M., Mukherjee, B.: Data evacuation from data centers in disaster-affected regions through software-defined satellite networks. Comput. Netw. **148**, 88–100 (2019)
17. Mart, R., Pardalos, P., Resende, M.: Handbook of Heuristics, 1st edn. Springer, Heidelberg (2018)
18. Martins, E., Pascoal, M.: A new implementation of Yen's ranking loopless paths algorithm. Q. J. Belgian French Italian Oper. Res. Soc. **1**(2), 121–133 (2003)
19. Nesmachnow, S.: Evaluating simple metaheuristics for the generalized Steiner problem. J. Comput. Sci. Technol. **5**(4) (2005)
20. Pulsipher, J.L., Zavala, V.M.: Measuring and optimizing system reliability: a stochastic programming approach. TOP 1–20 (2020)
21. Resende, M., Ribeiro, C.: Optimization by GRASP. Springer, Heidelberg (2016). https://doi.org/10.1007/978-1-4939-6530-4
22. Rodríguez-Pereira, J., Fernández, E., Laporte, G., Benavent, E., Martínez-Sykora, A.: The Steiner traveling salesman problem and its extensions. Eur. J. Oper. Res. **278**(2), 615–628 (2019)
23. Sartor, P., Robledo, F.: GRASP algorithms for the edge-survivable generalized Steiner problem. Int. J. Control Autom. **5**, 27–44 (2012)
24. Suzuki, H., Ishihata, M., Minato, S.: Designing survivable networks with zero-suppressed binary decision diagrams. In: Rahman, M.S., Sadakane, K., Sung, W.-K. (eds.) WALCOM 2020. LNCS, vol. 12049, pp. 273–285. Springer, Cham (2020). https://doi.org/10.1007/978-3-030-39881-1_23

# Max-Diversity Orthogonal Regrouping of MBA Students Using a GRASP/VND Heuristic

Matías Banchero[1], Franco Robledo[1], Pablo Romero[1(✉)], Pablo Sartor[2], and Camilo Servetti[1]

[1] Instituto de Computación, INCO, Facultad de Ingeniería,
Universidad de la República, Montevideo, Uruguay
{matias.banchero,frobledo,promero,camilo.servetti}@fing.edu.uy
[2] IEEM Business School, Universidad de Montevideo, Lord Ponsomby 2542,
Montevideo, Uruguay
psartor@um.edu.uy

**Abstract.** Students from Master in Business Administration (MBA) programs are usually split into teams. Many schools rotate the teams at the beginning of every term, so that each student works with a different set of peers during every term. Diversity within every team is desirable regarding gender, major, age and other criteria. Achieving diverse teams while avoiding -or minimizing- the repetition of student pairs is a time-consuming complex task for MBA Directors.

The Max-Diversity Orthogonal Regrouping (MDOR) problem is here introduced, where the goal is to maximize a global notion of diversity, considering multiple stages (i.e., terms) and intra-diversity within the teams. A hybrid GRASP/VND heuristic combined with Tabu Search is developed for its resolution. Its effectiveness has been tested in real-life groups from the MBA program offered at IEEM Business School, Universidad de Montevideo, Uruguay, with a notorious gain regarding team diversity and repetition level.

**Keywords:** MBA teams · Orthogonal regrouping · Diversity · GRASP · VND

## 1 Motivation

The collaborative team-formation and staffing/scheduling problems in workforce management is of paramount importance in projects deployment and large/scale corporations. Given the intrinsic hardness of multidisciplinary team-formation and clustering techniques, it is necessary to develop tools for this task. In this work we are focused on a maximum diversity regrouping assignment of MBA students; nevertheless, the reader can find potential applications in similar clustering problems. Experience shows that the student skills and learning process benefit significantly from highly-diverse teams when regarding prior experience,

© Springer Nature Switzerland AG 2021
N. Mladenovic et al. (Eds.): ICVNS 2021, LNCS 12559, pp. 58–70, 2021.
https://doi.org/10.1007/978-3-030-69625-2_5

age, gender, major and other features. MBA programs are usually split into four to six terms. Many MBA rotate the groups in every term so that students train their ability to adapt to different groups, benefit from new points of view and expand their peer network. Creating highly-diverse teams while keeping at a minimum the repetition of peer-pairs between terms is a very challenging problem faced by program directors at the beginning of every trimester.

The contributions of this paper can be summarized in the following items:

1. A novel combinatorial optimization problem called Max-Diversity Orthogonal Regrouping (MDOR) is here introduced. The goal is to find as many clusterings as terms, maximizing cluster diversity while keeping at a minimum the repetitions of pairs.
2. A GRASP/VND methodology combined with Tabu Search is developed.
3. The effectiveness of our proposal is tested with real-life students from the MBA program offered at IEEM Business School, Universidad de Montevideo, Uruguay.

The document is organized in the following manner. The related work is presented in Sect. 2. A mathematical programming formulation for the MDOR is introduced in Sect. 3. A full GRASP/VND heuristic combined with Tabu Search is presented in Sect. 4. Computational results based on real-life students are presented in Sect. 5. Section 6 contains concluding remarks and trends for future work.

## 2   Related Work

We identify the closest works of ours from the scientific literature in [2,3,7]. A simplified model with a large similarity in the team formation is presented in [3], which considers the dining philosophers problem for the assignment of students into groups. In [7], the problem is modeled using integer linear programming. This work considers a centroid for each cluster. Two approaches are studied: the min-sum approach tries to minimize the distances with respect to the centroid; the second is a min-max approach whose goal is to minimize the maximum (i.e., the worst) distance.

The case-study in [2] consists of the assignment of 235 students to 8 advisors. This work considers integer linear programming, and it is equivalent to the min-sum approach given by [7]. The problem belongs to the $\mathcal{NP}$-Hard class, and heuristics are available to tackle it [10]. A hybrid Genetic Algorithm is proposed in [9]. There, the authors suggest Tabu Search combined with strategic oscilations. Independently, [12] proposed an artificial bee-workers approach. In [8], a competitive General Variable Neighborhood Search (GVNS) is also proposed. An extension of this GVNS is offered in [4], with a Skewed VNS combined with a Shaking process to better explore the search-space. The goal in the Orthogonal Regrouping Problem is to partition a given set repeatedly, in such a way that every pair is included only once in some cluster. Well known instances have been extensively treated, e.g., the Kirkman's Schoolgirl Problem and the Social Golfer Problem.

Here we introduce the MDOR problem, which is suitable to the assignment of MBA students to teams that are re-built in every term. It is worth to remark that our approach has potential applications to other scenarios, such as staffing and scheduling in workforce management [5], team formation models for collaboration [14], and team-formation algorithms for faultline minimization [1], among others.

## 3    Problem

In this section, we describe the main features of our problem, and then we present a mathematical programming formulation. A brief discussion covers particular cases, which will be considered to address the problem heuristically.

### 3.1    Problem Description

Our problem formulation requires a definition of distance between any two items. In the context of grouping MBA students, the distance between two students would represent how different they are in terms of a set of criteria (age, type of major, gender, work experience, admission test score, etc.) that the MBA Director chooses. In the case of the real-life sets used in our test, the criteria are:

- Career (subdivided in percentage of Social Sciences, Natural and Exact Sciences content).
- Score in the Admission Test.
- Residence (urban or countryside).
- Gender.
- Age.

Career is split into three attributes in $[0, 1]$ which account for the relative levels of Social Sciences, Natural and Exact Sciences. The score in the Admission Test and the Age are natural numbers, while the remaining attributes assume binary domain. Once the attributes are selected, a distance function between the different individuals $d_{ij}$ must be specified. In what follows, the normalized-Euclidean distance is considered:

$$d_{ij} = d(x^i, x^j) = \frac{\|x^i - x^j\|_2}{max_{u \neq v}\|u - v\|_2}, \tag{1}$$

where the distance between each pair of students is found by a numerical assignment to the different attributes (i.e., different coordinates). Observe that this normalization implies that $0 \leq d_{ij} \leq 1$ for all the pairs of students $i$ and $j$ with corresponding attributes $x^i$ and $x^j$.

## 3.2  Problem Formulation

Consider the following variables:

- $N$ the number of students.
- $G$ the number of teams (clusters).
- $K$ the number of attributes.
- $M$ the number of students per team: $M = \frac{N}{G}$ (if integer).
- $S$ the number of terms (clusterings).
- $d_{ij}$ the distance between the students $i$ and $j$.
- $R$ is the number of terms that any pair of students can share ($R = 1$ for a SGP instance).

Consider the set of binary decision variables $x_{igs}$, such that $x_{igs} = 1$ if and only if the student $i$ is assigned to the group $g$ in term $s$, and $x_{igs} = 0$ otherwise. We introduce the MDOR problem as the following Integer Quadratic Problem:

$$\max_{x_{igs}} \sum_{s=1}^{S} \sum_{g=1}^{G} \sum_{i=1}^{N-1} \sum_{j=i+1}^{N} d_{ij} x_{igs} x_{jgs}, \tag{2}$$

$$s.t. \sum_{g=1}^{G} x_{igs} = 1, \; \forall (i,s) \in \{1,\ldots,N\} \times \{1,\ldots,S\} \tag{3}$$

$$\sum_{i=1}^{N} x_{igs} = M, \; \forall (g,s) \in \{1,\ldots,G\} \times \{1,\ldots,S\} \tag{4}$$

$$\sum_{s=1}^{S} \sum_{g=1}^{G} \sum_{i=1}^{N-1} \sum_{j=i+1}^{N} x_{igs} x_{jgs} \leq R, \; \forall (g,s) \in \{1,\ldots,G\} \times \{1,\ldots,S\} \tag{5}$$

$$x_{igs} \in \{0,1\}, \forall (i,g,s) \in \{1,\ldots,N\} \times \{1,\ldots,G\} \times \{1,\ldots,S\} \tag{6}$$

The goal is to maximize the diversity-sum among all clusters and clusterings, where the intra-cluster diversity is precisely the distance-sum among all the pairs of that cluster. Constraint 3 states that each student is included in a single team. Constraint 4 states that the teams have precisely $M$ students. Constraint 5 limits the number of times any pair of students can meet in different terms. Finally, Constraint 6 defines the binary domain for the decision variables.

## 3.3  Discussion

Observe that the previous MDOR model is adequate when $M = \frac{N}{G}$ is an integer. Next we comment on how to overcome this limitation and to minimize the number of repetitions as well.

**Number of Students per Group.** If $M = \frac{N}{G}$ is not an integer, we can replace Constraints 4 with a minimal variation. In fact, consider the Euclidean division: $N = G \times M + r$ for some remainder $r : 0 \leq r < G$. We can arrange $M + 1$ students in $r$ groups, and $M$ students in the remaining $G - r$ groups.

As a more general setting, pick two vectors $a$ and $b$ representing lower and upper-bounds on the number of students per group. Replace Constraints 4 with:

$$\sum_{g=1}^{G} x_{igs} \geq a_g, \; \forall(g,s) \in \{1, \ldots, G\} \times \{1, \ldots, S\}$$

$$\sum_{g=1}^{G} x_{igs} \leq b_g, \; \forall(g,s) \in \{1, \ldots, G\} \times \{1, \ldots, S\}.$$

**Avoiding Repetitions.** Avoiding repetitions is not always possible, depending on the parameters $G, M, S$ of a MDOR instance. Even when it is possible, no polynomial-complexity algorithm is known for the general case; variations like the SGP-completion problem are known to be NP-complete [6,13].

Let us consider a certain student, and let $w_s$ be the number of feasible peer students for him/her during the term $s$. The sequence $w_s$ satisfies the following recurrence:

$$w_1 = N - 1;$$
$$w_{i+1} = w_i - (M - 1),$$

since $M - 1$ new students are met in the last term $s = i$. A straight solution of the recurrence leads to $w_s = N - 1 - (s - 1)(M - 1)$. When the courses are finished we get $s = S$ and $w_S = N - 1 - (S - 1)(M - 1)$. Hence, if $N < (S - 1)(M - 1) + 1$, it is impossible to avoid repetitions.

Two possible heuristic approaches arise to cope with the repetition problem. One might build high-diversity solutions while controlling the repetition level. Alternative, one might generate repetition-free solutions and then choose and/or modify them seeking for improved diversity. In this paper we introduce an algorithm that follows the first approach. A parameter $GLOBAL\_REP$ is set; once more than $GLOBAL\_REP$ times a solution is generated including a repetition for a certain pair, the algorithm accepts the repetition.

## 4   Solution

GRASP and VND are well known metaheuristics that have been successfully used to solve many hard combinatorial optimization problems. GRASP is a powerful multi-start process which operates in two phases. A feasible solution is built in a first phase, whose neighborhood is then explored in the Local Search Phase. The second phase is usually enriched by means of different variable neighborhood structures. For instance, VND explores several neighborhood structures in a deterministic order. Its success is based on the simple fact that different neighborhood structures do not usually have the same local minimum. Thus, the resulting solution is simultaneously a locally optimum solution under all the neighborhood structures. The reader is invited to consult the comprehensive Handbook of Heuristics for further information [11]. Here, we develop a GRASP/VND methodology.

## 4.1    GRASP/VND Methodology for the MDOR

We followed a traditional VND flow diagram, that consists of three local searches:

- *Insert*: moves a student to another group.
- *Swap*: swaps two students from different groups.
- 3 − *Chain*: exchanges three students from three different groups.

The most simple local searches appear at the beginning. Therefore, the order is respectively *Insert*, *Swap* and 3 − *Chain*. A greedy randomized *Construction* phase takes effect first.

To speed-up the evaluation of the objective function, the internal structures in the main algorithm consider two vectors:

- $x^c[i]$: current group for student $i$, and
- $sd^c[i][g]$: current sum-diversity between the student $i$ and his/her peers in group $g$.

Observe that $sd^c[i][g] = \sum_{j:x[j]=g} d_{i,j}$, and if we link the students in a graph with link-weights $d_{i,j}$, by Handshaking Lemma we get that the objective is:

$$f(x^c) = \frac{1}{2} \sum_{i=1}^{N} sd^c[i][x^c[i]]. \tag{7}$$

In the following, the details of the construction and local searches are presented, in the respective order.

## 4.2    Construction Phase

The search space is the set of all student assignments to the groups, where each student belongs to exactly one group. A feasible solution also meets the respective lower and upper bounds $a_g$ and $b_g$. In our *Construction* phase, an iterative student insertion into groups takes effect, meeting the lower bounds $a_g$. Finally, in order to fulfill feasibility, all the students are assigned in some group, meeting the upper-bound $b_g$. Two factors are considered for these group-insertions: diversity and repetitions. In this construction phase, the priority is given to repetitions. Therefore, a memory with the previous terms is used, and if two assignment have identical number of repetitions, the assignment with maximum diversity is chosen. During the process, the diversity per group $g$ for some student $x$ is found using the following expression:

$$d'(x, g) = \sum_{y \in g} \frac{d(x, y)}{|g|}.$$

Observe the relation with the cardinality $|g|$; otherwise, groups with larger number of students are always preferred (Fig. 1).

---

**Algorithm 1** $Construction(studentGroup, a, b, atrsStandard, repMatrix)$

1: $studentVector \leftarrow \{1, 2, .., N\}$
2: $groupVector \leftarrow \{1, 2, .., N\}$
3: $assignOneRandomStudentToEachGroup(studentGroup, repMatrix)$
4: **while** $groupVector \neq \{\}$ **do**
5:     $selGroup \leftarrow assignGroupToStudForMinRepetitions($
6:     $studentGroup, repMatrix)$
7:     **if** $groupCount[selGroup] = a[selGroup]$ **then**
8:         $groupVector \leftarrow groupVector - selGroup$
9:     **end if**
10: **end while**
11: **for** $g \leftarrow 1$ $to$ $G$ **do**
12:     **if** $groupCount[g] = b[g]$ **then**
13:         $groupVector \leftarrow groupVector - g$
14:     **end if**
15: **end for**
16: **while** $groupVector \neq \{\}$ **do**
17:     $selGroup \leftarrow assignGroupToStudentForMinRepetitions($
18:     $studentGroup, repMatrix)$
19:     **if** $groupCount[selGroup] = b[selGroup]$ **then**
20:         $groupVector \leftarrow groupVector - selGroup$
21:     **end if**
22: **end while**

---

**Fig. 1.** Construction phase

The following variables are considered during the *Construction* phase:

- $studentGroup[s]$: the group assigned to student $s \in \{1, \ldots, N\}$.
- $atrsStandard[i, j]$: the value of attribute $j \in \{1, \ldots, K\}$ for the student $i$.
- $groupCount[g]$: the number of students in the group $g \in \{1, \ldots, G\}$.

The following functions are also considered:

- $assignOneRandomStudentToEachGroup()$: assigns, in each group, one random student uniformly picked at random.
- $assignGroupToStudForMinRepetitions()$: picks a random student, and assigns him/her to the group that leads to the least number of repetitions. Ties are solved using the maximum diversity.

## 4.3  Insertion

In this local search, a student $i$ is moved from a different group. We remark that a local search takes place whenever the resulting solution is both better and feasible. To test feasibility, we just check the lower and upper bounds for the old and the new group, respectively. The difference in the objective is the change in the diversity:

$$f(x^n) - f(x^c) = sd^c[i][g_2] - sd^c[i][g_1],$$

being $x^n$ the new solution and $x^c$ the current solution (Fig. 2).

---

**Algorithm 2** $Insertion(studentGroup, sd, solCurrent, atrsStandard, tabuMatrix)$

---

1:  $res \leftarrow false$
2:  **for** $i \leftarrow 1$ $to$ $N$ **do**
3:   **for** $g \leftarrow 1$ $to$ $G$ **do**
4:    **if** $studentGroup[i] \neq g$
5:    **and** $groupCount[g] < b[g]$ **and**
6:    $groupCount[studentGroup[i]] > a[g]$ **then**
7:     $diffSol \leftarrow sd[i][g] - sd[i][studentGroup[i]]$
8:     **if** $diffSol > 0$ **and** $updateTabuSearchMatrix($
9:     $i, g, studentGroup, tabuMatrix)$ **then**
10:      $studentGroup[i] \leftarrow g$
11:      $solCurrent \leftarrow solCurrent + diffSol$
12:      $updateSD(studentGroup, sd, i, g)$
13:      $res \leftarrow true$
14:     **end if**
15:    **end if**
16:   **end for**
17:  **end for**
18:  **return** $res$

---

**Fig. 2.** Local Search I: *Insertion*

## 4.4  Swap

In this local search, two students $i$ and $j$, originally belonging to different groups $g_i \neq g_j$, are exchanged, and the difference in the objective is:

$$f(x^n) - f(x^c) = (sd^c[i][g_j] - sd^c[i][g_i]) + (sd^c[j][g_j] - sd^c[j][g_i]) - 2d_{ij}$$

A pseudocode for *Swap* is presented in Fig. 3.

**Algorithm 3** $Swap(studentGroup, sd, solCurrent, atrsStandard, tabuMatrix)$

1:  $res \leftarrow false$
2:  **for** $i \leftarrow 1$ $to$ $N$ **do**
3:      **for** $j \leftarrow 1$ $to$ $N$ **do**
4:          **if** $studentGroup[i] \neq studentGroup[j]$ **then**
5:              $diffSol \leftarrow sd[i][studentGroup[j]] + sd[j][studentGroup[i]]$
6:              $-sd[i][studentGroup[i]] - sd[j][studentGroup[j]] - 2d_{i,j}$
7:              **if** $diffSol > 0$
8:              **and** $updateTabuSearchMatrix($
9:              $i, studentGroup[j], studentGroup, tabuMatrix)$
10:             **and** $updateTabuSearchMatrix($
11:             $j, studentGroup[i], studentGroup, tabuMatrix)$ **then**
12:                 $oldI \leftarrow studentGroup[i]$
13:                 $oldJ \leftarrow studentGroup[j]$
14:                 $studentGroup[i] \leftarrow oldJ$
15:                 $studentGroup[j] \leftarrow oldI$
16:                 $updateSD(studentGroup, sd, i, studentGroup[i])$
17:                 $updateSD(studentGroup, sd, j, studentGroup[j])$
18:                 $solCurrent \leftarrow solCurrent + diffSol$
19:                 $res \leftarrow true$
20:             **end if**
21:         **end if**
22:     **end for**
23: **end for**
24: **return** $res$

**Fig. 3.** Local Search II: $Swap$

### 4.5   3-Chain

Consider three different students $i$, $j$ y $k$ belonging to three different groups $g_i$, $g_j$ and $g_k$. Student $i$ is moved to $g_j$, $j$ is moved to $g_k$ and $k$ is moved to $g_i$ (Fig. 4):

$$f(x^n) - f(x^c) = (sd^c[i][g_j] - sd^c[i][g_i]) + (sd^c[j][g_k] - sd^c[j][g_j]) + (sd^c[k][g_i] - sd^c[k][g_k])$$
$$- (d_{ij} + d_{jk} + d_{ki})$$

### 4.6   Shake

In order to increase the diversity in the search-space, a shake process takes place. Consider a $k$-neighborhood of $Swap$ operation, this is, an arbitrary application of $k$ swaps. $Shake$ picks a $k$-neighbor, and the VND phase is re-started with the obtained solution, provided that the Tabu List allows for the shake to be done (i.e., controlling the repetitions threshold). Figure 5 presents a full pseudocode for $Shake$. In the general algorithm, $k$ starts equal to a parameter $K\_MIN$ and is increased by a second parameter $K\_STEP$ until the solution is improved or up to a third parameter $K\_MAX$.

---

**Algorithm 4** $3-Chain(studentGroup, sd, solCurrent, atrsStandard, tabuMatrix)$

1: $res \leftarrow false$
2: **for** $i \leftarrow 1$ $to$ $N$ **do**
3:     **for** $j \leftarrow 1$ $to$ $N$ **do**
4:         **for** $k \leftarrow 1$ $to$ $N$ **do**
5:             **if** $studentGroup[i] \neq studentGroup[j]$
6:             **and** $studentGroup[j] \neq studentGroup[k]$ **then**
7:                 $diffSol \leftarrow sd[i][studentGroup[j]] + sd[][studentGroup[k]]$
8:                 $+sd[k][studentGroup[i]] - sd[i][studentGroup[i]]$
9:                 $-sd[j][studentGroup[j]] - sd[k][studentGroup[k]]$
10:                $-2d_{i,j} - 2d_{j,k} - 2d_{k,i}$
11:             **if** $diffSol > 0$
12:             **and** $updateTabuSearchMatrix($
13:             $i, studentGroup[j], studentGroup, tabuMatrix)$
14:             **and** $updateTabuSearchMatrix($
15:             $j, studentGroup[k], studentGroup, tabuMatrix)$
16:             **and** $updateTabuSearchMatrix($
17:             $k, studentGroup[i], studentGroup, tabuMatrix)$ **then**
18:                 $oldI \leftarrow studentGroup[i]$
19:                 $oldJ \leftarrow studentGroup[j]$
20:                 $oldK \leftarrow studentGroup[k]$
21:                 $studentGroup[i] \leftarrow oldJ$
22:                 $studentGroup[j] \leftarrow oldK$
23:                 $studentGroup[k] \leftarrow oldI$
24:                 $updateSD(studentGroup, sd, i, studentGroup[i])$
25:                 $updateSD(studentGroup, sd, j, studentGroup[j])$
26:                 $updateSD(studentGroup, sd, k, studentGroup[k])$
27:                 $solCurrent \leftarrow solCurrent + diffSol$
28:                 $res \leftarrow true$
29:             **end if**
30:         **end if**
31:         **end for**
32:     **end for**
33: **end for**
34: **return** $res$

---

**Fig. 4.** Local Search III: $3 - Chain$

## 4.7  Main Algorithm

The main algorithm iterates over all terms. For each one, it starts by invoking $Construction$ a number of times $MAX\_TRIES$ that acts as a parameter. The most diverse solution is passed to the following step, where the following cycle is repeated a number of times $T\_MAX$ (another parameter): $Shake - Insertion - Swap - 3 - Chain$. The best solution found (the most diverse clustering) is chosen for the term, moving on to the next one.

---

**Algorithm 5** $Shake(studentGroup, k, sd, solCurrent, atrsStandard, tabuMatrix)$

1: **while** $k > 0$ **do**
2:    $randomI \leftarrow getRandom(N)$
3:    $randomJ \leftarrow getRandom(N)$
4:    **if** $studentGroup[randomI] <> studentGroup[randomJ]$ **then**
5:      **if** $updateTabuSearchMatrix(randomI,$
6:      $studentGroup[randomJ], studentGroup, tabuMatrix)$**and**
7:      $updateTabuSearchMatrix(randomJ,$
8:      $studentGroup[randomI], studentGroup, tabuMatrix)$ **then**
9:        $oldI \leftarrow studentGroup[i]$
10:       $oldJ \leftarrow studentGroup[j]$
11:       $studentGroup[i] \leftarrow oldJ$
12:       $studentGroup[j] \leftarrow oldI$
13:       $updateSD(studentGroup, sd, i, studentGroup[i])$
14:       $updateSD(studentGroup, sd, j, studentGroup[j])$
15:       $k \leftarrow k - 1$
16:     **end if**
17:   **end if**
18: **end while**
19: $updateSolCurrent(solCurrent, sd, studentGroup$
20: **return** $res$

---

**Fig. 5.** Perturbation Step: *Shake*

## 5   Computational Results

We carried out a comparison between the algorithm here introduced and the manual team assignment that was done in real-life with two IEEM Business School MBA cohorts from 2014 and 2015: "MBA1314" (34 students, 6 teams) and "MBA1415" (45 students, 8 teams).

The algorithm was coded in C++ and executed in a home-PC (Intel-core i7 2.2GHz, 8GB RAM). One hundred independent iterations were run (since GRASP is a multi-start metaheuristic) and the best solution was finally returned. As a preliminary stage, an adjustment of all the parameters was performed running several experiments. $MAX\_TRIES$ and $T\_MAX$ were set to 100 and 500 respectively. The *Shake* parameters were finally set to $K\_MIN = K\_STEP = 1$ and $K\_MAX = 3$. There is a trade-off between diversity and number of repetitions. A larger freezing-factor $GLOBAL\_REP$ in the Tabu List implies a lower level of diversity as one test with MBA1415 shows in Table 1. All results next reported were obtained with Tabu-list parameter to a freezing factor of 285.000 to keep repetitions at a minimum level.

Table 2 compares the diversity achieved by our algorithm vs the manual team assignment for the two cohorts and the five terms that the program spans; Table 3 does a similar comparison for repetitions per term. Our algorithm consistently outperformed the manual assignment when considering diversity and repetitions. It also took less time, since the longest execution took 50 min, while the manual assignment was reported to take more than 4 hours for each cohort.

**Table 1.** Diversity and repetitions per term, MBA1415: manual vs algorithm.

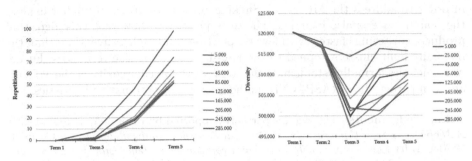

**Table 2.** Diversity per term, MBA1314 and MBA1415: manual vs algorithm.

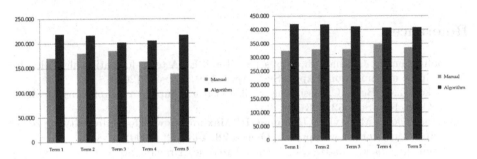

**Table 3.** Repetitions per term, MBA1314 and MBA1415: manual vs algorithm.

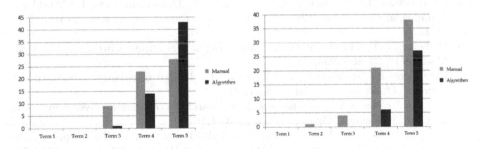

# 6   Conclusions and Trends for Future Work

A novel combinatorial optimization problem is introduced named Max-Diversity Orthogonal Regrouping (MDOR). It was conceived to cope with the problem of partitioning MBA cohorts into high-diversity teams, rotating the teams in every term and keeping under a given (low) threshold the repetitions. Nevertheless, the MDOR has potential applications in workforce management or team formation models for collaboration. The mathematical programming formulation is similar to a quadratic assignment problem, and the MDOR is presumably hard, even though a formal proof is not available in the literature.

A GRASP/VND methodology enriched with Tabu Search is here proposed in order to address the MDOR. A Shaking process in order to further explore the search-space is also included. The tests presented show that this algorithm produces clusterings faster, with fewer repetitions and higher diversities than the manually-built clusters applied to the real-life cohorts of the test cases. Future work includes formally establishing the computational complexity of the MDOR, and comparing our GRASP/VND methodology with alternative heuristics.

**Acknowledgements.** This work is partially supported by Project ANII FCE_1_2019 _1_156693 *Teoría y Construcción de Redes de Máxima Confiabilidad*, MATHAMSUD 19-MATH-03 *Rare events analysis in multi-component systems with dependent components* and STIC-AMSUD ACCON *Algorithms for the capacity crunch problem in optical networks*.

# References

1. Bahargam, S., Golshan, B., Lappas, T., Terzi, E.: A team-formation algorithm for faultline minimization. Expert Syst. Appl. **119**, 441–455 (2019)
2. Baker, B.M., Benn, C.: Assigning pupils to tutor groups in a comprehensive school. J. Oper. Res. Soc. **52**, 623–629 (2001)
3. Bhadurya, J., Mightyb, E.J., Damar, H.: Maximizing workforce diversity in project teams: a network flow approach. Omega **28**, 143–153 (2000)
4. Brimberg, Jack, Mladenovic, Nenad, Uroševic, Dragan: Solving the maximally diverse grouping problem by skewed general variable neighborhood search. Inf. Sci. **295**, 650–675 (2015)
5. De Bruecker, P., Van den Bergh, J., Beliën, J., Demeulemeester, E.: Workforce planning incorporating skills: state of the art. Eur. J. Oper. Res. **243**(1), 1–16 (2015)
6. Colbourn, C.J.: The complexity of completing partial Latin squares. Discret. Appl. Math. **8**(1), 25–30 (1984)
7. Desrosiers, J., Mladenović, N., Villeneuve, D.: Design of balanced MBA student teams. J. Oper. Res. Soc. **56**, 60–66 (2005)
8. Dragan, Uroševic: Variable neighborhood search for maximum diverse grouping problem. Yugoslav J. Oper. Res. **24**, 21–33 (2014)
9. Fan, Z.P., Chen, Y., Ma, J., Zeng, S.: A hybrid genetic algorithmic approach to the maximally diverse grouping problem. J. Oper. Res. Soc. **62**, 1423–1430 (2011)
10. Feo, T.A., Khellaf, M.: A class of bounded approximation algorithms for graph partitioning. Networks **20**, 181–195 (1990)
11. Mart, R., Pardalos, P.M., Mauricio, G., Resende, C.: Handbook of Heuristics, 1st edn. Springer, Heidelberg (2018). https://doi.org/10.1007/978-3-319-07124-4
12. Rodriguez, F.J., Lozano, M., García-Martínez, C., González, J.D.: An artificial bee colony algorithm for the maximally diverse grouping problem. Inf. Sci. **230**, 183–196 (2013)
13. Triska, M.: Solution methods for the social golfer problem. Inf. Sci. **295** (2008)
14. Wi, H., Seungjin, O., Mun, J., Jung, M.: A team formation model based on knowledge and collaboration. Expert Syst. Appl. **36**(5), 9121–9134 (2009)

# Scheduling in Parallel Machines with Two Servers: The Restrictive Case

Rachid Benmansour[1] and Angelo Sifaleras[2]([envelope])

[1] SI2M Laboratory, National Institute of Statistics and Applied Economics (INSEA), Rabat, Morocco
r.benmansour@insea.ac.ma
[2] Department of Applied Informatics, School of Information Sciences, University of Macedonia, 156 Egnatia Street, 54636 Thessaloniki, Greece
sifalera@uom.gr

**Abstract.** In this paper we study the Parallel machine scheduling problem with Two Servers in the Restrictive case (PTSR). Before its processing, the job must be loaded on a common loading server. After a machine completes processing one job, an unloading server is needed to remove the job from the machine. During the loading, respectively the unloading, operation, both the machine and the loading, respectively the unloading, server are occupied. The objective function involves the minimization of the makespan. A Mixed Integer Linear Programming (MILP) model is proposed for the solution of this difficult problem. Due to the NP-hardness of the problem, a Variable Neighborhood Search (VNS) algorithm is proposed. The proposed VNS algorithm is compared against a state-of-the-art solver using a randomly generated data set. The results indicate that, the obtained solutions computed in a short amount of CPU time are of high quality. Specifically, the VNS solution approach outperformed IBM CPLEX Optimizer for instances with 15 and 20 jobs.

**Keywords:** Scheduling · Parallel machine · Mixed integer programming · Variable neighborhood search · Single server

## 1 Introduction

Sequencing and scheduling decisions are crucial in manufacturing and service industries. Scheduling jobs on parallel machines consists of determining the starting time of each job and the machine that will process this job. The problem has a myriad of applications in logistics, manufacturing, and network computing etc. The literature on this subject is abundant as it is for the problem of parallel machine scheduling problem with single server [1,2,6,12].

Formally the problem of minimizing the makespan $C_{max}$ on the parallel machine with a single server is denoted by $Pm, S1|p_i, s_i|C_{max}$. In this notation, $m$ represents the number of machines, $S1$ represents the server, and $p_i$, $s_i$

© Springer Nature Switzerland AG 2021
N. Mladenovic et al. (Eds.): ICVNS 2021, LNCS 12559, pp. 71–82, 2021.
https://doi.org/10.1007/978-3-030-69625-2_6

represent the processing time and setup time (or loading time) of job $i$, respectively.

Considering this problem, Liu et al. [13] studied the objective of minimizing the total weighted job completion time. They proposed a branch-and-bound algorithm, a lower bound, and dominance properties.

In [12] the authors proposed two mixed integer programming formulations for the problem with several machines. They proposed also a hybrid heuristic algorithm combining Simulated Annealing (SA) and Tabu Search (TS) to minimize the total server waiting time. Recently El Idrissi et al. [7] proposed two additional mixed integer programming formulations for the same problem with better performance given especially by the time-indexed variables formulation.

In the paper of Torjai and Kruzslicz [17], the authors consider the situation where the shared server is used to unload the jobs. As an application, the authors studied a biomass truck scheduling problem. The trucks represent the machines in charge of delivering biomass from different locations to a single refinery operating a single server.

The problem of considering both loading and unloading operations in order to minimize the makespan was studied by Jiang et al. [11]. In their work the authors considered only two parallel machine and a single server that is capable of doing the loading and unloading operations. In addition to these assumptions, the authors considered the non-preemptive case in which the loading and unloading durations are both equal to one time unit. Given the NP-hardness nature of the problem, they applied a List Scheduling (LS) and Longest Processing Time (LPT) to solve this problem. Also they showed that LPT and LS have tight worst-case ratios of $4/3$ and $12/7$, respectively.

Some authors considered the problem with several servers. Ou et al. [15] studied the parallel machine scheduling with multiple unloading servers in order to minimize the total completion time of the jobs. As an application the authors cite milk run operations of a logistics company that faces limited unloading docks at the warehouse. The authors show that, the shortest-processing-time-first (SPT) algorithm has a worst-case bound of 2. They also provide heuristics and a branch-and-bound algorithm to solve the problem.

In [18] the authors studied the problem of scheduling non preemptively a set of jobs on identical parallel machines. Each job has to be loaded, on a given machine, by one of multiple servers available. The authors show that, the problem $Pm, Sk|s_i = 1|C_{max}$ is binary NP-hard and that the problem $Pm, S(m-1)|s_i = 1|C_{max}$ can be solved by a pseudo-polynomial algorithm. For a fixed number of machines and servers the problem is unary NP-hard when considering maximum lateness minimization.

As seen above, there are several articles dealing with the problem of parallel machine scheduling with loading and unloading operations. In some cases only one server was considered, and in other cases several servers were considered. Also several researchers considered only the loading or unloading operations whereas other researchers considered that the server can do both the loading operation and the unloading operation.

To the best of our knowledge, the case where two servers are available has not been studied before. One server is dedicated only for loading jobs on the machines and the other server is dedicated for unloading them from the machines. This problem is NP-hard as it is more difficult that the special case $P2, S1|p_j, s_j|C_{max}$ which is NP-hard (cf. [4]). The research contributions of this work are summarized as follows:

- This paper considers the parallel machine scheduling problem with loading and unloading servers.
- The general problem with restrictive and non-preemptive case is considered for the first time.
- An efficient VNS algorithm is proposed to solve this problem.

The remaining of the paper is organized as follows. The description of the problem and an illustrative example are given in Sect. 2. Section 3 presents the mathematical formulation of the restrictive model. Section 4 provides a VNS method for the solution of the PTSR. Finally, Sects. 5 and 6 present our experimental results and draw some conclusions, respectively.

## 2    Problem Description

This paper considers the parallel machine scheduling problem with loading and unloading servers. In this problem, we consider two machines and a set $\Omega_1 = \{1, 2, \ldots, n\}$ of $n$ independent jobs with integer processing times have to be processed non-preemptively on a set of parallel machines with two servers. The first server is dedicated to the loading of jobs on the machines and the second server realizes the unloading of jobs immediately after their execution. During the loading (respectively unloading) operation, both the machine and the loading server (respectively unloading server) are occupied and after this operation, the server becomes available for loading (respectively unloading) the next job. It is assumed that, the jobs are simultaneously available for processing at the beginning of the scheduling horizon, and that their processing times are fixed and known in advance. The objective function in the PTSR problem consists of minimizing the makespan.

### 2.1    Numerical Example

Let's assume we are given an instance with two parallel machines M1 and M2, eight jobs, and two servers. One server is used to load the jobs on one of the machines (denoted as **L**) and the other is used to unload them (denoted as **U**). The other data are displayed in the following Table 1. For each job $i$, $p_i$, $l_i$, and $u_i$ represent the processing time, the loading time, and the unloading time of this job, respectively.

The optimal objective function value can be obtained by solving the MIP formulation described in Sect. 3. IBM CPLEX Optimizer v12.6 requires 11.82 min to solve this problem. Figure 1 represents the corresponding schedule.

**Table 1.** Example instance for $n = 8$.

| Job | 1 | 2 | 3 | 4 | 5 | 6 | 7 | 8 |
|-----|---|---|----|---|---|---|---|---|
| $p_i$ | 6 | 5 | 10 | 7 | 7 | 9 | 6 | 5 |
| $l_i$ | 4 | 2 | 3 | 1 | 1 | 2 | 1 | 3 |
| $u_i$ | 3 | 1 | 1 | 1 | 2 | 3 | 4 | 3 |

In this schedule the server **L** loads first the job $J_7$ on the machine $M1$. This operation take one unit of time. Then the processing of this job begins from time $t = 1$ until $t = 7$. At this time the second server **U** is used to unload the $J_7$ from machine M1. This operation takes four units of time. The makespan in this schedule is equal to 46.

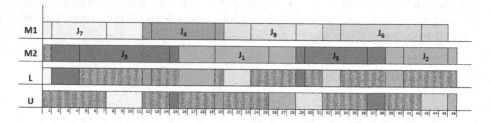

**Fig. 1.** The optimal schedule for the $n = 10$ problem instance

The problem is called *restrictive* because:

- once the job is loaded on the machine by the server **L**, it must be processed immediately by the machine.
- once the job is processed by the machine, it must be unloaded immediately by the server **U**.

If one of these conditions is not met, then the value of the optimal solution can be improved. In this case, a non-restrictive version of the problem occurs.

## 3   Mathematical Formulation of the Restrictive Model

In this case, the unloading of a job is carried out immediately after the end of its processing.
   Notations:

- $n$ the number of jobs
- $M = \{1, 2, ..., m\}$ set of the machines
- $p_i$ the processing time of job $i$
- $l_i$ the loading time of job $i$

- $u_i$ the unloading time of job $i$
- $B$ A large positive integer
- $\Omega_1 = \{1, 2, ..., n\}$ set of jobs to be processed on the machines
- $\Omega_2 = \{n+1, ..., 2n\}$ set of jobs to be processed on the loading server, each job $i$ has a duration $l_i$
- $\Omega_3 = \{2n+1, ..., 3n\}$ set of jobs to be processed on the unloading server, each job $i$ has a duration $u_i$
- $\Omega = \{1, 2, ..., 3n\}$ set of all the jobs

For the needs of modeling, we adopt the following notations: the $\rho$ parameter will represent the duration of jobs, whether on the machine or on the servers. Thus:

$$\forall i \in \Omega_1 \quad \rho_i = p_i + l_i + u_i$$

$$\forall i \in \Omega_2 \quad \rho_i = l_{i-n}$$

$$\forall i \in \Omega_3 \quad \rho_i = u_{i-2n}$$

**Variables:**
$C_i$ : the completion time of the job $i$
$x_{ik} = 1$ if job $i \in \Omega_1$ is processed on machine $k$ and 0 otherwise
$z_{ij} = 1$ if job $i$ is processed before job $j$ and 0 otherwise.

$$
\begin{aligned}
& min \ C_{max} && (1) \\
& s.t. \ C_{max} \geq C_i, \quad \forall i \in \Omega_1, && (2) \\
& \sum_{k=1}^{m} x_{ik} = 1, \quad \forall i \in \Omega_1, && (3) \\
& C_i \geq \rho_i, \quad \forall i \in J && (4) \\
& C_i \leq C_j - \rho_j + B(3 - x_{ik} - x_{jk} - z_{ij}), \quad \forall i \neq j \in \Omega_1, k \in M && (5) \\
& C_i \leq C_j - \rho_j + B(1 - z_{ij}), \quad \forall i \neq j \in \Omega_2 && (6) \\
& C_i \leq C_j - \rho_j + B(1 - z_{ij}), \quad \forall i \neq j \in \Omega_3 && (7) \\
& z_{ij} + z_{ji} = 1, \quad \forall i \neq j \in \Omega && (8) \\
& C_i - (p_i + u_i) = C_{i+n}, \quad \forall i \in \Omega_1 && (9) \\
& C_i = C_{i+2n}, \quad \forall i \in \Omega_1 && (10) \\
& x_{ij}, z_{ik} \in \{0, 1\}, \quad \forall i, j \in \Omega, \ j > i && (11)
\end{aligned}
$$

In this MIP model we aim to minimize the makespan $C_{max}$ (1). Constraints set (2) states that makespan of an optimal schedule is greater than or equal to the completion time of all executed jobs. In turns, the completion time of each job is at least greater than or equal to the duration of this job (4). Constraints (3) state that each job must be processed on exactly one machine. Constraints sets (5), (6) and (7) guarantee, respectively, that all jobs are scheduled on the loader, on the machines, and on the unloader without overlapping. The next constraints (8) impose that for each couple of jobs $(i, j)$, one must be processed before the other. Next, constraints (9) are used to calculate the completion time of each job $i$. Since we are dealing with the restrictive case, which states each job is immediately unloaded from the machine after its execution, then the completion time of the job, $C_i$, is equal to the completion time of the loading operation, $C_{i+n}$, plus the processing time and the unloading time $p_i + u_i$. Finally, constraints (10) are added to state that the completion time of the job $i$ on the machine is equal to the completion time of unloading operation of the same job. Constraint sets (11) define variables $x_{i,j}$ and $z_{i,k}$ as binaries.

# 4  Variable Neighborhood Search

## 4.1  General Variable Neighborhood Search

Variable Neighborhood Search (VNS) is a metaheuristic method based on systematic changes in the neighborhood structure initially proposed by Mladenović and Hansen [14]. The simplicity and efficiency of the VNS method has attracted several researchers the last decades and has lead to a large number of successful applications in a wide range of areas [3,16].

In this paper, we use the General VNS (GVNS) variant to solve the problem in hand. GVNS employs the Variable Neighborhood Descent (VND) which consists of a powerful local search step in each neighborhood rather than a simple local search step in only one neighborhood per iteration. Thus, the VND method constitutes the intensification part of the VNS and it is analytically described in [5]. The pseudo-code of the algorithm is presented below (Algorithm 1).

---

**Algorithm 1: GVNS**

---

**Data:** $x$, $l_{max}$, $k_{max}$, $t_{max}$
**Result:** Solution $x$
Generate an initial solution $x$;
**repeat**
    $l \leftarrow 1$ ;
    **repeat**
        $x' \leftarrow$ **Shake**$(x, k_{max}, l_{max})$;
        $x'' \leftarrow$ **VND**$(x', l_{max})$;
        $x, l \leftarrow$ **NeighborhoodChange**$(x, x'', l)$;
    **until** $l = l_{max}$;
    $t \leftarrow$ CpuTime() ;
**until** $t > t_{max}$;

---

The method **NeighborhoodChange**$(x, x'', l)$ is used to change (or not) the current neighborhood structure. If the local optimum $x''$ is better than the incumbent $x$, then **NeighborhoodChange** keeps this solution instead of $x$ (i.e.

$x \leftarrow x''$), and the search returns to $\mathcal{N}_1$; otherwise, it sets $l \leftarrow l+1$ in order to find, if possible, a better solution in a different neighborhood.

VND (Algorithm 2) starts with an initial solution $x_0$ and continuously tries to construct a new improved solution from the current solution $x$ by exploring its neighborhood $\mathcal{N}_l(x)$. The process continues to generate neighboring solutions until no further improvement can be made. In our implementation, we use the first improvement search strategy as we choose a random solution as the initial solution [9].

---

**Algorithm 2:** VND method

---

**Data:** $x, l_{max}$
**Result:** Solution $x$
repeat
    $l \leftarrow 1$;
    repeat
        Select $x' \in \mathcal{N}_l(x)$ such that $f(x') < f(x)$;
        **NeighborhoodChange**($x,x',l$);
    until $l = l_{max}$;
until *no improvement is made*;
return $x$

---

The aim of a Shaking procedure used within a VNS algorithm is to escape from local minima traps; thus, the Shaking method constitutes the diversification part of the VNS. The Shaking procedure performs a number of random jumps ($k \in \{1, 2, \ldots, k_{max}\}$) in a neighborhood $\mathcal{N}_l(x)$, $l \in \{1, 2, \ldots, l_{max}\}$ from a predefined set of neighborhoods. Note that, the $l$ index is given as input by Algorithm 1. In this work, $k_{max}$ was set equal to five based on some preliminary experiments.

---

**Algorithm 3:** Shake method

---

**Data:** $x, k_{max}, l_{max}$
**Result:** Solution $x$
for $k = 1$ to $k_{max}$ do
    Select randomly $x' \in \mathcal{N}_l(x)$;
    $x \leftarrow x'$;
end
return $x$

---

VNS uses a finite set of neighborhood structures denoted as $\mathcal{N}_l$, where $l \in \{1, 2, \ldots, l_{max}\}$. The $l^{th}$ neighborhood of solution $x$, $\mathcal{N}_l(x)$, is a subset of the search space, which is obtained from the solution $x$ by small changes. The VNS (Algorithm 1) includes an improvement phase in which a VND method is applied and a shaking phase used to escape local minima traps. These procedures are executed alternately until fulfilling a predefined stopping criterion. The stopping criterion of the proposed solution methodology was a maximum CPU time allowed for the VNS, equal to five minutes.

## 4.2  Neighborhood Structures

To design an efficient VNS algorithm one must carefully select the neighborhoods structure to use. Some authors recommend the use of less than three neighborhood structures [8]. We have developed the following three neighborhood structures ($l_{max} = 3$) for the computational experiments:

– Neighborhood $\mathcal{N}_1(x) = Swap(x)$: The neighborhood set consists of all permutations that can be obtained by swapping two adjacent jobs in the solution $x$.
– Neighborhood $\mathcal{N}_2(x) = Swap2(x)$: It consists of all solutions obtained from the solution $x$ swapping two random jobs.
– Neighborhood $\mathcal{N}_3(x) = Reverse(x)$: Given two jobs $j$ and $k$ we reverse the order of jobs being between those two jobs.

## 4.3  Initial Solution

The initial solution is chosen as a random permutation of the jobs. From any sequence of the jobs, we can build the solution as follows: We start by loading the first job on the machine $M_1$, and the second one on machine $M_2$. In this case, the second job is directly loaded after the end of loading job 1 (i.e., without idle time). For each one of the following jobs $j$, as soon as the server $\mathbf{L}$ becomes available, we can load job $j$ on one of the two available machines for processing. Otherwise, one should wait for one of the machines to be available before loading this job. It should be noted that, each time it must be checked that the end date of the unloading of job $j$ does not overlap with another job which is being unloaded. If it is the case, it is necessary to shift the starting time of loading job $j$ adequately.

## 4.4  Evaluation Function

Consider a permutation of jobs $\sigma = \{1, 2, \ldots, n\}$. To evaluate the value of the solution corresponding to $\sigma$ we will proceed as follows. At $t = 0$ all resources are available. The job 1 is scheduled on machine 1. This means that the job is loaded on $\mathbf{L}$ which will take $l_1$ units of time. The machine $M1$ that is busy up to this point will start the processing of this job. At the end of this operation, the job is unloaded immediately from the machine using the resource $\mathbf{U}$ ). Then we will schedule job 2 as soon as possible on machine 2. We may face two possibilities here. The first case is i) we will start loading this job on $\mathbf{L}$ just after the end of loading operation of job 1. In this case, the resource $\mathbf{U}$ is available when job 2 is to be unloaded (i.e. job 1 has finished unloading). The second case is ii) we will postpone the loading of job 2 so that at the time of unloading the resource $U$ will be available. For the following jobs, we must choose the earliest starting date on the server and on one of the two machines so as to have an execution without idle time and the smallest completion time possible on the resource $U$. Finally, the value of the solution $\sigma$ will be the completion time of the last scheduled job.

## 5 Computational Results

We generated the data as suggested by Hasani et al. [10]. Hence, we randomly generated server load $\eta$ in the interval $\{0.5, 1, 2\}$ for each server, where $\eta = E(s_i)/E(p_i)$ and $E(x)$ denotes the mean of $x$, and $s_i$ can either represents the loading time $l_i$ for the server $L$ or the unloading time $u_i$ for the server $U$. The processing times $p_j$ were uniformly distributed in the interval $(0, 100)$, and the loading and unloading times, respectively $l_j$ and $u_j$ were uniformly distributed in the interval $(0, 100\eta)$. Furthermore, we generated instances for $n \in \{15, 20\}$. Ten instances were randomly generated for each of the above values of $\eta$ and for the additional values of $n$.

All tests presented in this section were conducted on a personal computer running Windows 7 with an Intel®Core(™) i7 vPro with a clock speed at 2.90 GHz CPU and 16 GB of RAM. Also, IBM CPLEX Optimizer v12.6 was used for the solution of the MIP optimization problems.

In Table 2, which is subdivided into two parts, we have reported the results of 60 instances solved by VNS and by CPLEX. These cases relate to problems of size $n = 15$ jobs and problems of size $n = 20$ jobs. For $n = 15$, the first column represents the instance $k$. The second represents the values of the server load $\eta$. The third column represents the best value found by VNS in a time limit of five minutes. The fourth column represents the best value (upper bound) found by CPLEX in one hour. Finally, the last column, represents the relative MIP gap (difference between the lower and upper bounds) computed by CPLEX.

In Table 2, for instances with 15 jobs, VNS finds a better solution than CPLEX in 87% of the cases. VNS performance is even better for large instances. In fact, for $n = 20$, VNS always finds a better solution for each case than CPLEX, whether the time limit is five minutes or five seconds. We report here only the solutions found in five minutes since they were better than those found in five seconds. Note that in both cases, we have written in bold the best values found by VNS. Finally, we should highlight the fact that, the MIP model for $n = 20$, $m = 2$, $\eta = 0.5$ was not able to solve optimally the first instance even after six hours.

## 6 Conclusions and Future Work

This work studied the parallel machine scheduling problem with two servers in the restrictive case. Also, a mixed integer linear programming model and a variable neighborhood search approach were presented for the solution of the

**Table 2.** Computational results

| $n = 15$ | | | | | $n = 20$ | | | | |
|---|---|---|---|---|---|---|---|---|---|
| $k$ | $\eta$ | GVNS | CPLEX | GAP (%) | $k$ | $\eta$ | GVNS | CPLEX | GAP (%) |
| 1 | 0.5 | **818** | 818 | 45.35 | 1 | 0.5 | **1055** | 2342 | 88.95 |
| 2 | 0.5 | **873** | 873 | 39.09 | 2 | 0.5 | **1156** | 2450 | 89.14 |
| 3 | 0.5 | 784 | 782 | 43.48 | 3 | 0.5 | **991** | 2466 | 89.45 |
| 4 | 0.5 | **845** | 845 | 39.43 | 4 | 0.5 | **1151** | 2496 | 89.51 |
| 5 | 0.5 | **785** | 785 | 33.20 | 5 | 0.5 | **1052** | 2616 | 89.77 |
| 6 | 0.5 | **816** | 816 | 34.80 | 6 | 0.5 | **987** | 2399 | 88.22 |
| 7 | 0.5 | 697 | 695 | 21.57 | 7 | 0.5 | **1140** | 2631 | 90.07 |
| 8 | 0.5 | **832** | 832 | 39.06 | 8 | 0.5 | **989** | 2755 | 88.85 |
| 9 | 0.5 | **724** | 724 | 35.90 | 9 | 0.5 | **1033** | 2522 | 88.90 |
| 10 | 0.5 | **723** | 723 | 29.05 | 10 | 0.5 | **957** | 2756 | 90.28 |
| 1 | 1 | **1053** | 1053 | 33.20 | 1 | 1 | **1486** | 3897 | 89.09 |
| 2 | 1 | 1178 | 1177 | 41.83 | 2 | 1 | **1643** | 3801 | 87.57 |
| 3 | 1 | **1063** | 1064 | 28.20 | 3 | 1 | **1646** | 3548 | 87.78 |
| 4 | 1 | **1075** | 1079 | 32.43 | 4 | 1 | **1526** | 4090 | 90.56 |
| 5 | 1 | **1197** | 1198 | 39.65 | 5 | 1 | **1456** | 3774 | 88.45 |
| 6 | 1 | **1007** | 1008 | 31.80 | 6 | 1 | **1601** | 3968 | 88.82 |
| 7 | 1 | **1280** | 1283 | 42.18 | 7 | 1 | **1377** | 4227 | 89.70 |
| 8 | 1 | **1394** | 1395 | 45.96 | 8 | 1 | **1660** | 3560 | 88.07 |
| 9 | 1 | 1171 | 1168 | 36.30 | 9 | 1 | **1448** | 4041 | 89.58 |
| 10 | 1 | **1324** | 1325 | 40.24 | 10 | 1 | **1336** | 4207 | 89.46 |
| 1 | 2 | **2215** | 2224 | 39.53 | 1 | 2 | **2610** | 6323 | 88.40 |
| 2 | 2 | **2042** | 2049 | 36.12 | 2 | 2 | **2959** | 6149 | 87.75 |
| 3 | 2 | **2003** | 2009 | 37.36 | 3 | 2 | **2323** | 7413 | 88.17 |
| 4 | 2 | **1920** | 1926 | 34.74 | 4 | 2 | **2828** | 6366 | 86.30 |
| 5 | 2 | **1771** | 1777 | 33.65 | 5 | 2 | **2465** | 6605 | 88.49 |
| 6 | 2 | **1786** | 1786 | 41.71 | 6 | 2 | **2270** | 6450 | 88.62 |
| 7 | 2 | **1908** | 1914 | 33.52 | 7 | 2 | **2373** | 6191 | 87.70 |
| 8 | 2 | **2058** | 2059 | 44.52 | 8 | 2 | **2839** | 6368 | 88.08 |
| 9 | 2 | **2027** | 2040 | 39.69 | 9 | 2 | **2520** | 6936 | 90.58 |
| 10 | 2 | **2206** | 2210 | 40.32 | 10 | 2 | **2727** | 7182 | 89.23 |

proposed problem. The VNS solution approach showed a very good computational performance and computed solutions of better quality than CPLEX for instances with 15 or 20 jobs. Studying problems with more than two machines consists an interesting research idea for future work. Also, a similar model corresponding to the non-restrictive case is also interesting as a generalization of the proposed model.

# References

1. Abdekhodaee, A.H., Wirth, A.: Scheduling parallel machines with a single server: some solvable cases and heuristics. Comput. Oper. Res. **29**(3), 295–315 (2002)
2. Bektur, G., Saraç, T.: A mathematical model and heuristic algorithms for an unrelated parallel machine scheduling problem with sequence-dependent setup times, machine eligibility restrictions and a common server. Comput. Oper. Res. **103**, 46–63 (2019)
3. Benmansour, R., Sifaleras, A., Mladenović, N. (eds.): Variable neighborhood search. LNCS, vol. 12010. Springer, Cham (2020). https://doi.org/10.1007/978-3-030-44932-2
4. Brucker, P., Dhaenens-Flipo, C., Knust, S., Kravchenko, S.A., Werner, F.: Complexity results for parallel machine problems with a single server. J. Sched. **5**(6), 429–457 (2002)
5. Duarte, A., Mladenović, N., Sánchez-Oro, J., Todosijević, R.: Variable neighborhood descent. In: Martí, R., Panos, P., Resende, M. (eds.) Handbook of Heuristics, pp. 1–27. Springer, Cham (2016). https://doi.org/10.1007/978-3-319-07153-4_9-1
6. Elidrissi, A., Benbrahim, M., Benmansour, R., Duvivier, D.: Variable neighborhood search for identical parallel machine scheduling problem with a single server. In: Benmansour, R., Sifaleras, A., Mladenović, N. (eds.) ICVNS 2019. LNCS, vol. 12010, pp. 112–125. Springer, Cham (2020). https://doi.org/10.1007/978-3-030-44932-2_8
7. Elidrissi, A., Benmansour, R., Benbrahim, M., Duvivier, D.: MIP formulations for identical parallel machine scheduling problem with single server. In: 2018 4th International Conference on Optimization and Applications (ICOA), pp. 1–6. IEEE (2018)
8. Glover, F.W., Kochenberger, G.A.: Handbook of Metaheuristics, vol. 57. Springer, Heidelberg (2006)
9. Hansen, P., Mladenović, N., Pérez, J.A.M.: Variable neighbourhood search: methods and applications. Ann. Oper. Res. **175**(1), 367–407 (2010). https://doi.org/10.1007/s10479-009-0657-6
10. Hasani, K., Kravchenko, S.A., Werner, F.: Block models for scheduling jobs on two parallel machines with a single server. Comput. Oper. Res. **41**, 94–97 (2014)
11. Jiang, Y., Zhang, Q., Hu, J., Dong, J., Ji, M.: Single-server parallel-machine scheduling with loading and unloading times. J. Comb. Optim. **30**(2), 201–213 (2014). https://doi.org/10.1007/s10878-014-9727-z
12. Kim, M.Y., Lee, Y.H.: MIP models and hybrid algorithm for minimizing the makespan of parallel machines scheduling problem with a single server. Comput. Oper. Res. **39**(11), 2457–2468 (2012)
13. Liu, G.S., Li, J.J., Yang, H.D., Huang, G.Q.: Approximate and branch-and-bound algorithms for the parallel machine scheduling problem with a single server. J. Oper. Res. Soc. **70**(9), 1554–1570 (2019)
14. Mladenović, N., Hansen, P.: Variable neighborhood search. Comput. Oper. Res. **24**(11), 1097–1100 (1997)
15. Ou, J., Qi, X., Lee, C.Y.: Parallel machine scheduling with multiple unloading servers. J. Sched. **13**(3), 213–226 (2010). https://doi.org/10.1007/s10951-009-0104-1
16. Sifaleras, A., Salhi, S., Brimberg, J. (eds.): Variable Neighborhood Search. LNCS, vol. 11328. Springer, Cham (2019). https://doi.org/10.1007/978-3-030-15843-9

17. Torjai, L., Kruzslicz, F.: Mixed integer programming formulations for the biomass truck scheduling problem. CEJOR **24**(3), 731–745 (2016). https://doi.org/10.1007/s10100-015-0395-6
18. Werner, F., Kravchenko, S.A.: Scheduling with multiple servers. Autom. Remote Control **71**(10), 2109–2121 (2010). https://doi.org/10.1134/S0005117910100103

# Reduced Variable Neighbourhood Search for the Generation of Controlled Circular Data

Sergio Consoli[1]([✉])[iD], Domenico Perrotta[1][iD], and Marco Turchi[2][iD]

[1] European Commission, Joint Research Centre, Via E. Fermi 2749,
21027 Ispra, VA, Italy
{sergio.consoli,domenico.perrotta}@ec.europa.eu
[2] Fondazione Bruno Kessler-IRST, Via Sommarive 18, 38123 Povo, Trento, Italy
turchi@fbk.eu

**Abstract.** A number of artificial intelligence and machine learning problems need to be formulated within a directional space, where classical Euclidean geometry does not apply or needs to be readjusted into the circle. This is typical, for example, in computational linguistics and natural language processing, where language models based on Bag-of-Words, Vector Space, or Word Embedding, are largely used for tasks like document classification, information retrieval and recommendation systems, among others. In these contexts, for assessing document clustering and outliers detection applications, it is often necessary to generate data with directional properties and units that follow some model assumptions and possibly form close groups. In the following we propose a Reduced Variable Neighbourhood Search heuristic which is used to generate high-dimensional data controlled by the desired properties aimed at representing several real-world contexts. The whole problem is formulated as a non-linear continuous optimization problem, and it is shown that the proposed Reduced Variable Neighbourhood Search is able to generate high-dimensional solutions to the problem in short computational time. A comparison with the state-of-the-art local search routine used to address this problem shows the greater efficiency of the approach presented here.

**Keywords:** Optimization · Reduced Variable Neighbourhood Search · Synthetic directional data generation · Circular statistics

## 1 Introduction and Background

The analysis and interpretation of directional data requires specific data representations, descriptions and distributions. Directional data occurs in many application areas like, e.g. earth sciences, astronomy, meteorology and medicine. Note that directional data is an "interval type" data: points are typically considered on the unit radius circle, or sphere, and the focus is on the direction of the

© The Author(s) 2021
N. Mladenovic et al. (Eds.): ICVNS 2021, LNCS 12559, pp. 83–98, 2021.
https://doi.org/10.1007/978-3-030-69625-2_7

data vectors. The position of the "zero degrees" is then arbitrary, and the angle between two unit length vectors is used as a natural measure for their distance. Although the angular distances can be used sometimes also with general data vectors defined in the Euclidean space, usual statistics, such as the arithmetic mean and the standard deviation, are not appropriate because they do not have this rotational invariance property [34], and one must rely on proper directional (or circular) statistics [26]. In this context classical Euclidean geometry laws do not apply or need to be properly reformulated into the circle.

Nowadays also a number of popular artificial intelligence and machine learning problems are modelled using directional statistics. This is typical, for example, in computational linguistics and natural language processing, where language models based on Bag-of-Words [42], Vector Space [35], or Word Embedding [25], are largely used for tasks like document classification, information retrieval and recommendation systems, among others. In these models, documents are represented by feature vectors. Specifically, in Bag-of-Words, a text (such as a sentence or a document) is represented as the bag (multiset) of its words, disregarding grammar and even word order but keeping multiplicity. A Vector Space model is a special Bag-of-Words representing a document by a feature vector, where each feature is a word (term) and the feature's value is a term weight. The term weight might be, either, a binary value (with 1 indicating that the term occurred in the document, and 0 indicating that it did not); a term frequency value (indicating how many times the term occurred in the document); or a TF-IDF (term frequency-inverse document frequency) value [1], a numerical statistic that is intended to reflect how important a word is to a document into a collection or corpus. The entire document is thus a feature vector, and each feature vector corresponds to a point in a vector space. The model for this vector space is such that there is an axis for every term in the vocabulary, and so the vector space is $v$-dimensional, where $v$ is the size of the vocabulary. For long documents, where the vocabulary can be excessively large to handle, Word Embedding models can be a preferred option [25]. In Word Embedding model, like e.g. the popular Word2Vec [27], GloVe [31], or the most recent BERT [18], words or phrases from the vocabulary are mapped to lower-dimensions vectors of real numbers. Conceptually it involves a mathematical embedding from a space with many dimensions per word to a continuous vector space with a much lower dimension [25]. Although these representation models have been largely used in computational linguistics and natural language processing, they have found also applications for generic visual categorization in computer vision [6,16,40], among others.

Typically, reasons for preferring a directional approach in these contexts are driven by the problem. For example, the angular distance is known to favor vector components with higher variance (see e.g. [33]), which in linguistics is desirable for giving more weight to the more informative terms. Other reasons are of computational nature. When the space dimension $v$ and sample size $n$ may be both very large, the computation of the ordinary distances may become time consuming. In addition, the high dimensional vectors are often very sparse and an angular distance, namely the cosine similarity, would neglect the 0-components and save time compared to an ordinary Euclidean distance. Besides, $v$ is often considerably larger than $n$, leading to singular covariance matrices and complicating the use of the Mahalanobis distance (see e.g. [20], [15] and [19]), which otherwise is in general a good choice for taking into account data correlations.

In cluster analysis, as in many statistical and data science problems, the distance measure plays a crucial role [8,10]. Points are assigned to groups according to their distance from the closer group centroid, and points that cannot be firmly assigned to any group are considered outliers [13]. A clustering method, e.g. K-means [36], may produce equivalent results with different distance measures (and centroid definitions) but behaves very differently with others.

To study the relationship between distance measures and objectively compare the performance of clustering, outliers detection, or other data analysis frameworks, it is often necessary to conduct systematic tests on a big number of different data sets artificially generated with known properties. In fact, it is well known that there is always some bias in the use of standard collections of data from real life applications[1] which should be rather used to confirm the properties of methods previously studied under controlled settings [12].

As an example, [38] have extended the Forward Search [2] method to the automatic detection of outliers in human label documents set. These outliers are documents that are wrongly assigned to a category or weakly correlated to other documents, and that need to be removed before training a learning system on these data. The synthetic datasets produced with our method allow to calibrate the outliers detection strategy and to assess the quality of the information retrieval process [38].

While it is rather easy to generate synthetic multivariate data according to some probability distribution or covariance structure in the Euclidean space, efficient methods that at the same time impose constraints on the distribution of the pairwise angular distances of the generated data vectors have been less explored in the literature. In [14] we have proposed a simple local search approach for addressing the problem of high-dimensional directional data generation controlled by specific properties. In particular, motivated by problems in computational linguistics, it is hypothesized that data are power-law distributed [19,34], and that their pairwise cosine similarities are distributed as a von Mises distribution [41], i.e. the analogous of the normal distribution in circular statistics [26], around a desired cosine value. This represents a common realistic setting in several computational linguistics applications [22,32].

In this contribution we propose an efficient optimization approach based on a Reduced Variable Neighbourhood Search (RVNS) heuristic [9,21,30] which is characterized by a very high speed and some interesting implementation improvements with respect to the local search approach in the literature [14]. This algorithm is characterized by an ease of implementation and simplicity, and it takes

---

[1] see for example the UC Irvine Machine Learning Repository (UCI): https://archive. ics.uci.edu/ml/index.php, a popular open-source repository of annotated datasets for machine learning tasks.

inspiration from the recently proposed "less is more approach" [10,29], which supports the adoption of non-sophisticated and effective metaheuristics instead of hard-to-reproduce and complex solution approaches. We will show how the proposed RVNS heuristic is able to generate efficiently high-dimensional directional data controlled by the desired properties. The generation algorithm is computationally efficient and flexible as well in terms of adaptability to other distributions and problems.

The rest of the paper is organized as follows. Section 2 formulated the whole problem as a non-linear continuous optimization problem. Section 3 describes the implementation details of the considered optimization algorithms, namely the state-of-the-art local search routine in the literature and our proposed RVNS approach. Our computational analysis is reported in Sect. 4, where it is shown that the proposed RVNS is able to generate high-dimensional solutions to the problem in shorter computational time with respect to the local search routine, demonstrating the efficiently of the approach presented here. Finally the paper ends with conclusions and suggestions for possible future research in Sect. 5.

## 2    Problem Formulation

The problem has been firstly formulated by [14] distinguishing between two general settings. The first is the ideal case of a single homogeneous group of $n$ documents represented in a vector space of size $v$ (vocabulary size). Here, high recall and precision are achieved by minimizing the sum of the pairwise similarities between documents,

$$F = \sum_{i=1}^{n} \sum_{j=1, j \neq i}^{n} cs(Z_i, Z_j) \tag{1}$$

where $cs$ is the cosine similarity function that for two vectors $Z_i$ and $Z_j$ is:

$$cs(Z_i, Z_j) = \frac{Z_i \cdot Z_j}{\|Z_i\|_2 \cdot \|Z_j\|_2} = \frac{\sum_{k=1}^{v} z_{(i,\,k)} \cdot z_{(j,\,k)}}{\sqrt{\sum_{k=1}^{v} z_{(i,\,k)}^2} \cdot \sqrt{\sum_{k=1}^{v} z_{(j,\,k)}^2}}. \tag{2}$$

Clearly, $0 \leq cs(Z_i, Z_j) \leq 1$. In the second setting we assume $K > 1$ homogeneous groups and a number of isolated outliers. Now, the quantity $F$ computed within a given homogeneous group should be maximized, while it should be minimized when computed on the set of the $K$ estimated centroids[2]. The "density" of the documents within an homogeneous group is determined by the average value of the quantity $F$ for that group. Therefore, to generate artificial data suitable for benchmarks in the two general settings we can use the following constrained non-linear continuous optimization problem formulation [24].

Given a cosine similarity value $\widetilde{cs} \in (0, 1)$, a tolerance $\xi \in (0, 1), \xi << \widetilde{cs}$, find a set $\mathscr{Z}$ of $n$ $v$-dimensional vectors with $v >> n$ non-negative variables, such

---

[2] It can be shown that the mean vector estimated within a group is a centroid with good properties (see e.g. [37], Section 8.2.6).

that the set of $n(n-1)/2$ pairwise cosine similarity values $\mathscr{C} = \{cs(Z_i, Z_j) \; ; \; j = 2, \cdots, n \; ; \; 1 < i < j\}$ satisfies:

1. $cs(Z_i, Z_j) \in [\widetilde{cs} - \xi, \; \widetilde{cs} + \xi]$,
2. $\mathscr{C}$ is a random sample drawn from a von Mises distribution

$$f(x|\mu, \kappa) = \frac{\exp^{\kappa(\cos(x-\mu))}}{2\pi I_0(\kappa)} \quad 0 < x \leq 2\pi$$

with mean $\mu(\mathscr{C}) \approx \widetilde{cs}$ and concentration parameter $\kappa(\mathscr{C}, \xi)$, being $I_0(\cdot)$ the modified Bessel function of order 0.

We use the Watson goodness-of-fit test [39] to assess whether $\mathscr{C}$ is consistent with the hypothesized von Mises null distribution with known mean ($\widetilde{cs}$) and concentration parameter estimated from $\mathscr{C}$ itself.

In the von Mises, the concentration parameter $\kappa$ is the analogous of the reciprocal of the variance for the normal distribution and links the sample standard deviation $\sigma(\mathscr{C})$ with the chosen tolerance $\xi$. This is done by using the well-known empiric *three-sigma rule* [34] and monitoring the constraint $\sigma(\mathscr{C}) \approx \xi/(10/3)$ during the data generation process, which corresponds to a concentration parameter $\kappa(\mathscr{C}, \xi) \approx 9$, ensuring that the data are representative of more than 99.9% of the generating von Mises distribution.

## 3   Optimization Methods

This section contains the implementation details of the considered optimization algorithms, namely the state-of-the-art local search routine implemented in [14] and the RVNS approach that we propose here. For a survey on the basic concepts of approximate optimization methods, including stochastic local search and metaheuristics, the reader is referred to [3,4,23]. The algorithms have been implemented using the commercial software package MATLAB, R2019b 64-bit, version 9.7.0 [3].

### 3.1   Local Search Algorithm

The local search algorithm starts from an initial solution, $\mathscr{X}^0$, of $n$ data vectors with $v$ variables, constructed as described in [14] such that all the vectors have exactly the same pairwise cosine similarity value $\widetilde{cs}$ [11]. Consider the set $\mathscr{X}$ of $n$ data vectors with $v \gg n$ variables, where all the vector tails are equal from variable index $n+1$ onwards, i.e. $z_{(i, k)} = z_{(j, k)} \; \forall k \geq n+1$ and $\forall i, j \leq n$, with $i \neq j$. For simplicity, all equal tail variables will be denoted without the vector index, e.g. $z_{(k)}$ for variable index $k \geq n+1$, where $z_{(k)} = z_{(i, k)} = z_{(j, k)}$ $\forall i, j \leq n, i \neq j$. Furthermore, the vectors share the same diagonal variable value, say $z_D$, i.e. $z_{(i, i)} = z_{(j, j)} = z_D \; \forall i, j \leq n$. The remaining variables of the vectors

---

[3] ©™The MathWorks Inc., Natick, MA (US). The source code of the algorithms is available to download upon request from the authors.

of $\mathscr{Z}$ are all set equal to a same constant value, say $z_C$, i.e. $z_{(i,\,k)} = z_{(j,\,k)} = z_C$ $\forall k \leq n$ and $\forall i, j \leq n$, with $i \neq j$. As proved in [14], all row vectors of such initial solution, referred to as $\mathscr{Z}^0$, have the same pairwise distance for geometric construction. This allows our local search algorithm to start from a solution which already satisfies the first constraint of the problem. Summarizing, the initial solution $\mathscr{Z}^0$ is [14]:

$$\mathscr{Z}^0 = \begin{pmatrix} Z_1 \\ Z_2 \\ \cdots \\ Z_{n-1} \\ Z_n \end{pmatrix} = \begin{pmatrix} z_D & z_C & \cdots & \cdots & z_C & z_{(n+1)} & \cdots & z_{(v)} \\ z_C & z_D & z_C & \cdots & \vdots & z_{(n+1)} & \cdots & z_{(v)} \\ \vdots & \ddots & \ddots & \ddots & \vdots & \vdots & \vdots & \vdots \\ \vdots & \vdots & z_C & z_D & z_C & z_{(n+1)} & \cdots & z_{(v)} \\ z_C & \cdots & \cdots & z_C & z_D & z_{(n+1)} & \cdots & z_{(v)} \end{pmatrix}. \quad (3)$$

Consider an arbitrary cosine similarity value $\widetilde{cs} \in (0,1)$, and let $z_C$, $z_{(n+1)}$, $\ldots$, $z_{(v)}$ be an arbitrary set of non-negative reals (i.e. $\in \Re^+$). Then, $\forall i, j \leq n$, with $i \neq j$, $cs(Z_i, Z_j) = \widetilde{cs}$ i.i.f.

$$z_D = \frac{z_C + \sqrt{z_C^2 - \widetilde{cs} \cdot \{z_C^2 + (\widetilde{cs} - 1)[(n-1)z_C^2 + \sum_{k=n+1}^{v} z_{(k)}^2]\}}}{\widetilde{cs}}. \quad (4)$$

Calculate $cs(Z_i, Z_j)$, $\forall Z_i, Z_j \in \mathscr{Z}^0$, directly by the definition of cosine similarity, and by replacing the component values with those specified in Eq. 3–4. By computing the resulting term, the expression is invariant $\forall i, j \leq n$, $i \neq j$, and it ends exactly in $\widetilde{cs}$. For more details on these calculations the reader is referred to [14].

It is worth noting in Eq. 4 that $\Delta = z_C^2 - \widetilde{cs} \cdot \{z_C^2 + (\widetilde{cs} - 1)[(n-1)z_C^2 + \sum_{k=n+1}^{v} z_{(k)}^2]\}$ is always non-negative. By computing directly $\Delta \geq 0$, trivially it is obtained:

$$z_C^2 \geq -\frac{\widetilde{cs} \cdot \sum_{k=n+1}^{v} z_{(k)}^2}{1 + (n-1)\widetilde{cs}}, \quad (5)$$

which is always true because: $\widetilde{cs} \geq 0$; $\sum_{k=n+1}^{v} z_{(k)}^2 > 0$; $(n-1)\widetilde{cs} > 0$ as $n$ is always $> 1$; and then, being the second member of the inequality a negative number, the first member, $z_C^2$, is always bigger than the second member [14].

Such produced initial solution $\mathscr{Z}^0$ represents a good starting point for the whole optimization algorithm because it already satisfies the first constraint of the problem (i.e. $cs(Z_i, Z_j) \in [\widetilde{cs} - \xi, \ \widetilde{cs} + \xi], \forall Z_i, Z_j \in \mathscr{Z}$). However $\mathscr{Z}^0$ violates the second problem constraint, i.e. that the distribution of the data in $\mathscr{C}$ is a random sample drawn from a von Mises distribution with mean $\mu(\mathscr{C}) \approx \widetilde{cs}$ and and concentration parameter $\kappa(\mathscr{C}, \xi) \approx 9$, i.e. $\sigma(\mathscr{C}) \approx \xi/(10/3)$. Indeed, by construction of the initial solution $\mathscr{Z}^0$, the distribution of the data in $\mathscr{C}$ is a Dirac centered in $\widetilde{cs}$.

The local search routine proceeds now by perturbing such initial solution $\mathscr{Z}^0$ in order to move gradually the distribution of the data in $\mathscr{C}$ from the Dirac to a von Mises distribution with the desired properties. This recurring disturbance is performed by an iterative process which, at each iteration, injects a noise $\mathcal{N}$ normally distributed, with $\mu = 0$, to the incumbent solution. The perturbation is greedily accepted iif the modified solution does not violate both the first and second constraints of the problem; otherwise the perturbation is rejected. The satisfaction of the second problem constraint is checked by performing a Watson goodness-of-fit test [39] at 1% significant level to ensure that the modified intermediate solution is consistent with the Von Mises distribution with the desired properties. The first problem constraint is validated instead by making sure that all pairwise cosine similarities among the perturbed vectors fall within the desired interval $[\widetilde{cs} - \xi, \ \widetilde{cs} + \xi]$. For this reason, it is needed to inject a noise being small enough to be very likely to be accepted, but not excessively small otherwise the local search routine would become to slow in terms of convergence within a reasonable number of steps. [14] found experimentally that a random noise $\mathcal{N}$ whose components follows a normal distribution with $\mu(\mathcal{N}) = 0$ and $\sigma(\mathcal{N}) = 0.01$ is a good choice yielding a satisfactory tradeoff between convergence speed and constraints violations given by the random perturbation.

The overall local search routine stops when it is reached a perturbed solution satisfying all problem constraints and having a standard deviation of the pairwise cosine similarities equal or larger than the desired threshold value of $\xi/(10/3)$ given by the second problem constraint.

## 3.2   Reduced Variable Neighbourhood Search Heuristic

An important drawback of the local search routine just described in the previous section is that the normal perturbation that is applied to the incumbent solution is static and does not adapt automatically to the size of the problem to handle. Although in [14] it has been fixed experimentally a random normal noise $\mathcal{N}$ having $\mu(\mathcal{N}) - 0$ and $\sigma(\mathcal{N}) - 0.01$ as a good perturbation setting, on average, it is very likely that, either, this $\sigma(\mathcal{N})$ value will be too small for large size problems, yielding to an excessively slow convergence time for the algorithm, or too large for small size problem instances, producing a large number of constraints violations which will increase as well exponentially the computational running time of the entire procedure.

For this reason we propose here an intelligent optimization approach based on Reduced Variable Neighbourhood Search (RVNS) [9,21,30] aimed at achieving high-quality performance for the problem by automating the resulting optimization strategy and adapting online to the size of the problem to tackle. The aim it to lead the local search to achieve a proper balance of diversification (exploration) and intensification (exploitation) during the search process, a fundamental objective for any effective heuristic solution approach. The diversification capability of a metaheuristic refers to its aptitude of exploring thoroughly different zones of the search space in order to identify promising areas. When a promising area is detected, the metaheuristic needs to exploit it intensively to find the relative local-optimum, but at the same time without wasting excessive computational resources. This is referred as the intensification capability of the metaheuristic.

Finding a good balance between diversification and intensification is indeed an essential task for the proper effectiveness of a metaheuristic [3,4,23].

RVNS is a variant of the classic Variable Neighbourhood Search (VNS) algorithm [5,10,21], which is a popular metaheuristic based on dynamic changes of neighbourhood structures of an incumbent solution during the search process [9,21]. The VNS methodology is based on the core concept of searching for new solutions in increasingly distant neighbourhoods of the current solution, jumping only if a better solution is found, without being limited only to a fixed trajectory [30]. RVNS is a variant that has been shown to be successful for many combinatorial problems where local optima with respect to one or several neighbourhoods are relatively close to each other [28].

The variable metric method has been suggested by [17]. The idea is to change the metric in each iteration such that the search direction (steepest descent with respect to the current metric) adapts better to the local shape of the function. The RVNS method is obtained if random points are selected from the current neighbourhood under exploration and no descent is followed. Rather, the values of these new points are compared with that of the incumbent and updating takes place in case of improvement. RVNS is a typical example of a pure stochastic heuristic, akin to a classic Monte-Carlo method, but more systematic [28]. It is useful especially for very large problem instances for which the local search within the classic VNS approach might become costly, as it is the case with our problem.

The details of our heuristic based on Reduced Variable Neighbourhood Search for the given problem are specified in Algorithm 1. The algorithm starts with an initial solution $\mathscr{L}$ equal to $\mathscr{L}^0$, that is the same starting solution in Eq. 3 described previously in Sect. 3.1. It is by construction a Dirac centered in $\widetilde{cs}$, already satisfying the first constraint of the problem but not the second.

Then, the *shaking phase*, which represents the core idea of RVNS, is applied to $\mathscr{L}$. The shaking phase aims to change the neighbourhood structure, $N_k(\cdot)$, of the incumbent solution to achieve a larger algorithm diversification. The new incumbent solution, $\mathscr{L}$, is generated at random in order to avoid cycling, which might occur if a deterministic rule is used.

The simplest and most common choice for the neighbourhood structure consists of setting neighbourhoods with increasing cardinality: $|N_1(\cdot)| < |N_2(\cdot)| < ... < |N_{k_{max}}(\cdot)|$, where $k_{max}$ represents the maximum size of the shaking phase. Let $k$ and $k_{step}$ be, respectively, the current size and the step size of the shaking phase.

The algorithm starts by selecting the first neighbourhood ($k \leftarrow 1$) and, at each iteration, it increases the parameter $k$ if the new incumbent solution violates one of the problem constraints ($k \leftarrow k + 1$), until the largest neighbourhood is reached ($k = k_{max}$). The process of changing neighbourhoods when no improvement occurs diversifies the search trajectory. In particular, the choice of neighbourhoods of increasing cardinality yields a progressive diversification of the search process. Note that the $k_{step}$ parameter has been introduced in order to adapt the classic RVNS schema from a combinatorial to a continuous optimization setting.

For the given problem, a shaking phase of size $k$ consists of perturbing the incumbent solution $\mathscr{X}$ with a random noise $\mathscr{N}$ normally distributed with $\mu = 0$ and $\sigma(\mathscr{N}) = k \cdot k_{step}$, producing a perturbed solution in the $N_k(\cdot)$ of the current solution. Note that each perturbation corresponds just to a limited local modification of the incumbent solution $\mathscr{X}$ which hopefully will not violate both

---

**Input:** The number of vectors to generate, $n$, with $v \gg n$ non-negative data variables; the reference cosine similarity value $\tilde{cs} \in (0, 1)$; and the tolerance $\xi \ll \tilde{cs}$ to bound the cosine similarities in $[\tilde{cs} - \xi, \tilde{cs} + \xi]$

**Output:** A set $\mathscr{X}$ of $n$ vectors with $v$ non-negative data variables;

*Initialization:*
- Let $\mathscr{C} = \bigcup_{i,j=1, j>i}^{n} cs(Z_i, Z_j)$ be the distribution of the pairwise cosine similarities of the vectors in $\mathscr{X}$;
- Let $k$, $k_{step}$, and $k_{max}$, respectively the current size, the step size, and the maximum size, of the shaking phase;

**begin**
    · Generate the initial solution $\mathscr{X}^0$ of $n$ vectors with $v$ variables:
    $(\mathscr{X}^0, \mathscr{C}) \leftarrow Initial\text{-}solution\text{-}construction(n, v, \tilde{cs})$;
    · Set $\mathscr{X} = \mathscr{X}^0$;
    · Set $k_{step}$ and $k_{max}$ arbitrarily;
    **repeat**
        Set $k \leftarrow 1$;
        **while** $k < k_{max}$ **do**
          · Set $\mathscr{X}' = \mathscr{X}$;
          · Select at random the noise matrix $\mathscr{N}^{n,v}$ whose components follows a normal distribution with $\mu(\mathscr{N}) = 0$ and $\sigma(\mathscr{N}) = k \cdot k_{step}$;
          · Perturb $\mathscr{X}$: $z_{(i, j)} = z_{(i, j)} \cdot (1 + \mathscr{N}_{(i, j)})$     $\forall i = 1, \ldots, n; \forall j = 1, \ldots, v$;
          · Evaluate the pairwise cosine similarities of the modified vectors in
          $\mathscr{X}$:  $\mathscr{C} = \bigcup_{i,j=1, j>i}^{n} cs(Z_i, Z_j)$;
          **foreach** $c \in \mathscr{C}$ **do**
              **if** $(c \notin [\tilde{cs} - \xi, \tilde{cs} + \xi])$ **then**   //first constraint violated
                 · Restore the previous solution: $\mathscr{X} = \mathscr{X}'$;
                 **if** $k == k_{max}$ **then**
                    · Halve the step size of the shaking phase $k_{step} \leftarrow k_{step}/2$;
                 **else**
                    · Increase the current size of the shaking phase: $k \leftarrow k + 1$;
                 **end**
                 · **break**;
              **end**
          **end**
          **if** $(\chi^2\text{-}test(\mathscr{C}, \tilde{cs})) == False)$ **then**   //second constraint violated
             · Restore the previous solution: $\mathscr{X} = \mathscr{X}'$;
             **if** $k == k_{max}$ **then**
                 · Halve the step size of the shaking phase $k_{step} \leftarrow k_{step}/2$;
             **else**
                · Increase the current size of the shaking phase: $k \leftarrow k + 1$;
             **end**
          **end**
          **if** $\mathscr{X} \neq \mathscr{X}'$ **then**   //new perturbed solution accepted
             · Double the step size of the shaking phase $k_{step} \leftarrow k_{step} \cdot 2$;
             · Restart with the first neighbourhood structure: $k \leftarrow 1$;
          **end**
        **end**
    **until** $\sigma(\mathscr{C}) \geq \xi/(10/3)$;
    $\Rightarrow Return(\mathscr{X})$.
**end**

**Algorithm 1:** Reduced Variable Neighbourhood Search heuristic

problem constraints, as already described for the basic local search procedure in Sect. 3.1. Otherwise, if one of the two constraints are violated by $\mathscr{Z}$, the new solution is discarded by restoring the previous solution ($\mathscr{Z} = \mathscr{Z}'$), and this is perturbed again in a larger neighbourhood ($k \leftarrow k + 1$).

The process of increasing progressively the parameter $k$ in case of no improvements occurs until the maximum size of the shaking phase, $k_{max}$, is reached. When this happens, meaning that the value of $k_{step}$ may be too large having produced already $k_{max}$ consecutive unsuccessful perturbations with the same $k_{step}$, the step size of the shaking phase is halved ($k_{step} \leftarrow k_{step}/2$) in order to produce next perturbations having standard deviation proportionally smaller ($\sigma(\mathscr{N}) = k \cdot k_{step}$). In this way when the algorithm restarts from the first neighbourhood ($k \leftarrow 1$) of $\mathscr{Z}$, a fine-grained noise will be iteratively produced such that to increase the acceptance chances of the next perturbations.

Conversely, if the perturbed solution $\mathscr{Z}$ does not violate both problem constraints (i.e. $\mathscr{Z} \neq \mathscr{Z}'$), this is accepted as the new incumbent solution and the search is restarted from its first neighbourhood ($k \leftarrow 1$). In this case, the value of $k_{step}$ is doubled ($k_{step} \leftarrow k_{step} \cdot 2$) in order to produce next perturbations having standard deviation proportionally larger. This simple reactive schema is aimed at achieving an optimal setting of $k_{step}$ in order to speeding up the converge speed of the algorithm. Given this reactive schema, the value of $k_{max}$ is not relevant to the overall performance of the algorithm, as this value depends directly from $k_{step}$ which is optimally tuned on-line. Therefore the value of $k_{max}$ is set equal to 5 at the beginning of the algorithm and is not required to change.

The algorithm continues iteratively with the same procedure and stops when the standard deviation of the pairwise cosine similarities of the obtained solution $\mathscr{Z}$ is equal or larger than the expected standard deviation, $\xi/(10/3)$, that is the prefixed problem goal, giving $\mathscr{Z}$ as the final output.

# 4   Computational Analysis

This section reports our computational experiments on the comparison of the performance of the proposed RVNS approach with respect to that of the local search routine in the literature. The heuristics are identified with the following abbreviations: *LS*, for the local search routine described in Sect. 3.1; and *RVNS*, for the Reduced Variable Neighbourhood Search implementation described in Sect. 3.2.

In the experiments we generate data vectors drawn from a power-law distribution [7][4], which is a realistic setting in computational linguistics [35], and considering a number of components, $v$, ranging from $1 \cdot 10^3$ to $100 \cdot 10^3$ components.

As shown in [14], the easiest setting for our problem is to fix the desired cosine similarity value, $\widetilde{cs}$ to 0.5. This setting is sufficient for comparing the performance of the two optimization algorithms by varying the number of components $v$ only.

---

[4] Although we generate data from a power-law distribution the approaches can be easily generalized to any kind of distribution of the data.

**Table 1.** Computational results of the compared algorithms ($LS$ and $RVNS$) for $\widetilde{cs} =$ 0.5 and $v$ ranging from $1 \cdot 10^3$ to $10 \cdot 10^3$ components. The reported results are the average values over 10 problem instances for each components dimension $v$. For each algorithm, the first column is the average computational running time in seconds (*time*); the second column is the average of the number of unsuccessful iterations (*unsucc*), that is when the perturbed solutions violated one of the problem constraints; the third column is the average of the total number of iterations required by the algorithm to stop (*tot iter*).

| size $v$ | $LS$ | | | $RVNS$ | | |
|---|---|---|---|---|---|---|
| | *time* | *unsucc* | *tot iter* | *time* | *unsucc* | *tot iter* |
| 1000 | 1.00 | 123.2 | 172.7 | 0.43 | 66.1 | 77.4 |
| 2000 | 2.21 | 114.9 | 162.2 | 1.41 | 80.8 | 94.3 |
| 3000 | 3.18 | 123.5 | 169.0 | 1.60 | 70.7 | 82.9 |
| 4000 | 3.86 | 104.8 | 154.7 | 2.40 | 84.3 | 98.8 |
| 5000 | 5.23 | 122.7 | 167.6 | 2.97 | 83.7 | 98.2 |
| 6000 | 6.67 | 125.4 | 176.9 | 4.98 | 118.4 | 137.4 |
| 7000 | 7.04 | 111.0 | 158.3 | 4.47 | 94.5 | 110.1 |
| 8000 | 9.29 | 126.2 | 180.4 | 3.40 | 61.5 | 71.7 |
| 9000 | 8.88 | 109.2 | 156.2 | 6.65 | 105.9 | 123.3 |
| 10000 | 10.29 | 106.1 | 149.1 | 6.67 | 88.6 | 103.8 |
| **Total:** | **57.65** | **1167.0** | **1647.1** | **34.98** | **854.5** | **997.9** |

Indeed the computational times to generate the vectors increase proportionally with the number of components $v$ of the vectors. In particular, for high values of $v$, the computational times may become quite large. In addition, the desired tolerance value, $\xi$, is set to 0.1, yielding to a bounded interval for the pairwise cosine similarities: $[\widetilde{cs} - 0.1, \widetilde{cs} + 0.1] = [0.4, 0.6]$ for each considered cosine similarity value $\widetilde{cs}$. The number of vectors to be generated is also an arbitrary user choice; in our study we have chosen to generate 100 vectors for each of the controlled circular data group, as in [14].

Our computational results are reported in Table 1 and Table 2, which report a comparison of $LS$ and $RVNS$ respectively for small a large instances of the problem. In particular, Table 1 contains the results obtained by the two algorithms by setting $\widetilde{cs} = 0.5$ and letting $v$ range from $1 \cdot 10^3$ to $10 \cdot 10^3$ number of components, with a step of $1 \cdot 10^3$. Table 2 contains instead the results obtained by the two algorithms for larger instances with $\widetilde{cs} = 0.5$ and $v$ ranging from $10 \cdot 10^3$ to $100 \cdot 10^3$, with a step of $10 \cdot 10^3$ components. In both tables, for each dataset having components dimension $v$ we have generated 10 different problem instances, therefore the results reported in the tables are the average values among the 10 generated problem instances[5]. For each algorithm, the first column

---

[5] All the computations have been made on an Intel Core i7 microprocessor at 2.6 GHz with 16.0 GB RAM.

**Table 2.** Computational results of the compared algorithms ($LS$ and $RVNS$) for larger problem instances with $\widetilde{cs} = 0.5$ and $v$ ranging from $10 \cdot 10^3$ to $100 \cdot 10^3$ components. The reported results are the average values over 10 problem instances for each components dimension $v$. For each algorithm, the first column is the average computational running time in seconds ($time$); the second column is the average of the number of unsuccessful iterations ($unsucc$), that is when the perturbed solutions violated one of the problem constraints; the third column is the average of the total number of iterations required by the algorithm to stop ($tot\ iter$).

| size $v$ | $LS$ | | | $RVNS$ | | |
|---|---|---|---|---|---|---|
| | time | unsucc | tot iter | time | unsucc | tot iter |
| $10 \cdot 10^3$ | 10.29 | 106.1 | 149.1 | 6.67 | 88.6 | 103.8 |
| $20 \cdot 10^3$ | 20.92 | 103.1 | 149.9 | 12.41 | 85.6 | 101.4 |
| $30 \cdot 10^3$ | 32.09 | 118.1 | 164.2 | 19.34 | 87.8 | 102.3 |
| $40 \cdot 10^3$ | 44.15 | 116.1 | 166.7 | 31.54 | 102.4 | 119.5 |
| $50 \cdot 10^3$ | 52.55 | 113.1 | 159.8 | 40.87 | 107.6 | 126.1 |
| $60 \cdot 10^3$ | 63.86 | 114.2 | 160.6 | 40.56 | 88.2 | 103.8 |
| $70 \cdot 10^3$ | 68.16 | 100.9 | 146.4 | 50.82 | 95.7 | 111.4 |
| $80 \cdot 10^3$ | 91.59 | 123.1 | 173.7 | 44.55 | 73.4 | 85.9 |
| $90 \cdot 10^3$ | 105.11 | 127.5 | 178.3 | 59.42 | 86.8 | 101.6 |
| $100 \cdot 10^3$ | 111.04 | 121.6 | 170.4 | 71.74 | 89 | 104.3 |
| **Total:** | **599.76** | **1143.8** | **1619.0** | **377.92** | **905.1** | **1060.1** |

in the tables is the average computational running time in seconds ($time$); the second column is the average of the number of unsuccessful iterations ($unsucc$), that is when the perturbed solutions violated one of the problem constraints; the third column is the average of the total number of iterations required by the algorithm to stop ($tot\ iter$).

Looking at Table 1 containing smaller problem instances, $RVNS$ performed better than $LS$ in all cases. It was able to generate the required controlled datasets in almost half of the time required by $LS$. The superiority of $RVNS$ with respect to $LS$ is also highlighted by the smaller number of unsuccessful perturbations, $unsucc$, and the smaller number of iterations, $tot\ iter$, obtained by $RVNS$. Practically the heuristic based on Reduced Variable Neighbourhood Search results to be a smarter optimization routine than the classic local search in the literature, being able to produce recurrent perturbations with a higher acceptance likelihood.

As shown in Table 2, these results are also confirmed for larger instances of the problem. $RVNS$ generated the desired controlled circular data in shorter computational time than $LS$, requiring an inferior number of total iterations and producing a smaller number of unsuccessful perturbations as well. As we can see from this table, for higher values of $v$ the computational times become quite large (see e.g. the case with $v = 100 \cdot 10^3$), confirming the high complexity of the problem. Nevertheless *RVNS* was able to generate the controlled data in considerably shorter computational time with respect to *LS*, confirming the superiority of the Reduced Variable Neighbourhood Search approach. As shown in our computational analysis *RVNS* is able to produce quickly circular data with the desired properties for vectors with very large components dimension $v$, which represents an important achievement.

Figure 1 shows an example of a circular dataset generated by the RVNS procedure. The histogram of the pairwise cosine similarities distribution in the figure has the typical shape of a normal distribution, which is in fact very similar in the circle to a von Mises of concentration parameter 9. The values fall within the expected interval $[\widetilde{cs} - 0.1, \ \widetilde{cs} + 0.1] = [0.4, \ 0.6]$, showing the ability of the algorithm to produce the data with the desired characteristics.

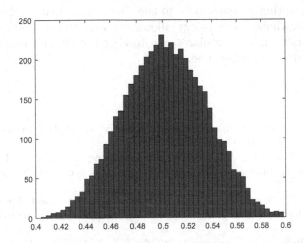

**Fig. 1.** An example of a circular dataset generated by the RVNS procedure with $\widetilde{cs} = 0.5$. As expected the histogram of the pairwise cosine similarities distribution for samples of the generated power-law data vectors has the shape of a von Mises distribution, i.e. the analogous of the normal distribution in the circle.

# 5   Conclusions

In many artificial intelligence and machine learning environments where stored entities are compared with each other or with incoming patterns, it is often necessary to compute artificially similar data to be used for experiments, simulations, or generally for test purposes. Hypothesizing a computational linguistic setting, we have modelled this problem as a constrained non-linear continuous optimization problem where the goal is to generate controlled datasets of similar circular vectors satisfying the conjecture in Eq. 1 for $n$ documents in a vector space of size $v$ (vocabulary size), and leaving the user the choice of the degree of closeness between the data (i.e. the setting of the pairwise similarity value among the generated vectors). Assuming a cosine similarity metric among the data, the problem has been conduced to the artificial generation of similar data in the circular space, further increasing the complexity of the problem. In addition, as these controlled datasets are aimed to represent real-world settings in computational linguistics, we have added the further constraint that such circular data need to be power-law distributed, while their pairwise cosine similarity values should follow a von Mises distribution within a desired bounded similarity interval.

In order to address the problem and to produce solutions within reasonable computational running time we have elaborated an optimization procedure based on Reduced Variable Neighborhood Search. In our computational experience we have shown the superiority of the proposed approach with respect to a local search routine in the literature in terms, in particular, of convergence speed. The proposed Reduced Variable Neighbourhood Search is efficient and scale well by adapting automatically to the dimension of the problem to tackle. The presented optimization strategy allows the generation of high-dimensional controlled datasets in the circular space having the desired properties within small computational running time.

# References

1. Aizawa, A.: An information-theoretic perspective of TF-IDF measures. Inf. Process. Manag. **39**(1), 45–65 (2003)
2. Atkinson, A.C., Riani, M., Cerioli, A.: Exploring Multivariate Data with the Forward Search. Springer, New York (2004). https://doi.org/10.1007/978-0-387-21840-3
3. Borenstein, Y., Moraglio, A.: Theory and Principled Methods for the Design of Metaheuristics. Springer, Heidelberg (2013). https://doi.org/10.1007/978-3-642-33206-7
4. Borne, P., Philip, F.G., Popescu, D., Stefanoiu, D., Kamel, A.E.: Optimization in Engineering Sciences: Approximate and Metaheuristic Methods. Wiley, New York (2014)
5. Braysy, O.: A reactive variable neighborhood search for the vehicle routing problem with time windows. INFORMS J. Comput. **15**(4), 347–368 (2003). https://doi.org/10.1287/ijoc.15.4.347.24896
6. Chum, O., Philbin, J., Zisserman, A.: Near duplicate image detection: min-hash and TF-IDF weighting. In: BMVC 2008 - Proceedings of the British Machine Vision Conference 2008 (2008)
7. Clauset, A., Shalizi, C.R., Newman, M.E.J.: Power-law distributions in empirical data. SIAM Rev. **51**(4), 661–703 (2009)
8. Consoli, S., Darby-Dowman, K., Geleijnse, G., Korst, J., Pauws, S.: Heuristic approaches for the quartet method of hierarchical clustering. IEEE Trans. Knowl. Data Eng. **22**(10), 1428–1443 (2010)
9. Consoli, S., Darby-Dowman, K., Mladenović, N., Moreno-Pérez, J.: Variable neighbourhood search for the minimum labelling Steiner tree problem. Ann. Oper. Res. **172**(1), 71–96 (2009). https://doi.org/10.1007/s10479-008-0507-y
10. Consoli, S., Korst, J., Pauws, S., Geleijnse, G.: Improved metaheuristics for the quartet method of hierarchical clustering. J. Global Optimiz. **78**(2), 241–270 (2020). https://doi.org/10.1007/s10898-019-00871-1

11. Consoli, S., Moreno Pérez, J.: Solving the minimum labelling spanning tree problem using hybrid local search. Electron. Notes Discrete Math. **39**, 75–82 (2012)
12. Consoli, S., Recupero, D.R., Petkovic, M. (eds.): Data Science for Healthcare: Methodologies and Applications. Springer, Cham (2019). https://doi.org/10.1007/978-3-030-05249-2
13. Consoli, S., Stilianakis, N.: A quartet method based on variable neighborhood search for biomedical literature extraction and clustering. Int. Trans. Oper. Res. **24**(3), 537–558 (2017)
14. Consoli, S., Turchi, M., Perrotta, D.: A local search approach for generating directional data. Intell. Data Anal. **20**(2), 439–453 (2016)
15. Croux, C., Gelper, S., Haesbroeck, G.: Robust scatter regularization. In: Saporta, G., Lechevallier, Y. (eds.) Compstat 2010, Book of Abstracts. Paris: Conservatoire National des Arts et Metiers (CNAM) and the French National Institute for Research in Computer Science and Control (INRIA), p. 138 (2010)
16. Csurka, G., Dance, C.R., Fan, L., Willamowski, J., Bray, C.: Visual categorization with bags of keypoints. In: Workshop on Statistical Learning in Computer Vision, ECCV, pp. 1–22 (2004)
17. Davidović, T.: Scheduling heuristic for dense task graphs. Yugoslav J. Oper. Res. **10**, 113–136 (2000)
18. Devlin, J., Chang, M.W., Lee, K., Toutanova, K.: BERT: Pre-training of Deep Bidirectional Transformers for Language Understanding. arXiv e-prints arXiv:1810.04805 (2018)
19. Filzmoser, P., Maronna, R., Werner, M.: Outlier identification in high dimensions. Comput. Stat. Data Anal. **52**(3), 1694–1711 (2008)
20. Friedman, J., Hastie, T., Tibshirani, R.: Sparse inverse covariance estimation with the graphical lasso. Biostatistics **9**(3), 432–441 (2008)
21. Hansen, P., Mladenović, N.: Variable neighborhood search. In: Marti, R., Pardalos, P., Resende, M. (eds.) Handbook of Heuristics, pp. 759–787. Springer, Heidelberg (2018). https://doi.org/10.1007/978-3-319-07124-4_19
22. Hoff, P.H.: Simulation of the matrix Bingham-von Mises-fisher distribution with application to multivariate and relational data. J. Comput. Graph. Stat. **18**, 438–456 (2009)
23. Hoos, H.H., Stutzle, T.: Stochastic Local Search: Foundations and Applications. Morgan Kaufmann Publishers, San Francisco (2005)
24. Jeyakumar, V., Rubinov, A.: Continuous Optimization: Current Trends and Modern Applications. Springer, Boston (2005). https://doi.org/10.1007/b137941
25. Kusner, M., Sun, Y., Kolkin, N., Weinberger, K.: From word embeddings to document distances. In: 32nd International Conference on Machine Learning, ICML 2015, vol. 2, pp. 957–966 (2015)
26. Mardia, K.V., Jupp, P.E.: Directional Statistics. Wiley, Hoboken (2008)
27. Mikolov, T., Sutskever, I., Chen, K., Corrado, G., Dean, J.: Distributed representations of words and phrases and their compositionality. In: Advances in Neural Information Processing Systems, NIPS 2013 (2013)
28. Mladenović, N., Petrović, J., Kovačević-Vujčić, V., Čangalović, M.: Solving spread spectrum radar polyphase code design problem by tabu search and variable neighbourhood search. Eur. J. Oper. Res. **151**(2), 389–399 (2003)
29. Mladenović, N., Todosijević, R., Urošević, D.: Less is more: basic variable neighborhood search for minimum differential dispersion problem. Inf. Sci. **326**, 160–171 (2016)

30. Pei, J., Darzić, Z., Drazić, M., Mladenović, N., Pardalos, P.: Continuous variable neighborhood search (C-VNS) for solving systems of nonlinear equations. INFORMS J. Comput. **31**, 235–250 (2019)
31. Pennington, J., Socher, R., Manning, C.: GloVe: global vectors for word representation. In: EMNLP 2014–2014 Conference on Empirical Methods in Natural Language Processing, Proceedings of the Conference, pp. 1532–1543 (2014)
32. Presutti, V., Nuzzolese, A., Consoli, S., Gangemi, A., Reforgiato Recupero, D.: From hyperlinks to semantic web properties using open knowledge extraction. Semant. Web **7**(4), 351–378 (2016)
33. Qian, G., Sural, S., Pramanik, S.: A comparative analysis of two distance measures in color image databases. In: Proceedings of the 2002 International Conference on Image Processing, vol. 1, pp. I-401–I-404 (2002)
34. Rice, J.: Mathematical Statistics and Data Analysis, 2nd edn. Duxbury Press, Pacific Grove (1995)
35. Salton, G., Wong, A., Yang, C.S.: A vector space model for automatic indexing. Commun. ACM **18**, 613–620 (1975)
36. Steinbach, M., Karypis, G., Kumar, V.: A comparison of document clustering techniques. In: KDD Workshop on Text Mining, pp. 420–428 (2000)
37. Tan, P.N., Steinbach, M., Kumar, V.: Introduction to Data Mining. Addison-Wesley, Boston (2006)
38. Turchi, M., Perrotta, D., Riani, M., Cerioli, A.: Robustness issues in text mining. In: Kruse, R., Berthold, M., Moewes, C., Gil, M., Grzegorzewski, P., Hryniewicz, O. (eds.) Synergies of Soft Computing and Statistics for Intelligent Data Analysis. AISC, vol. 190, pp. 263–272. Springer, Heidelberg (2013). https://doi.org/10.1007/978-3-642-33042-1_29
39. Watson, G.S.: Goodness-of-fit tests on a circle. Biometrika **48**(1/2), 109–114 (1961)
40. Weston, J., Bengio, S., Usunier, N.: Large scale image annotation: learning to rank with joint word-image embeddings. Mach. Learn. **81**(1), 21–35 (2010). https://doi.org/10.1007/s10994-010-5198-3
41. Wood, A.T.A.: Simulation of the von Mises Fisher distribution. Commun. Stat. - Simul. Comput. **23**(1), 157–164 (1994)
42. Zhang, Y., Jin, R., Zhou, Z.H.: Understanding bag-of-words model: a statistical framework. Int. J. Mach. Learn. Cybern. **1**(1–4), 43–52 (2010). https://doi.org/10.1007/s13042-010-0001-0

# Sequential and Parallel Scattered Variable Neighborhood Search for Solving Nurikabe

Paul Bass[✉] and Aise Zulal Sevkli[✉]

Denison University, Granville, OH 43023, USA
{bass_p1,sevklia}@denison.edu

**Abstract.** Japanese pencil games have been the subjects of innumerable papers. However, some problems - like Sudoku - receive far more attention than others - like Nurikabe. In this paper we propose a novel algorithm to solve Nurikabe puzzles. We first introduce a sequential hybrid algorithm that we call *Scattered Variable Neighborhood Search*. We then propose a method of parallelizing this algorithm, examining the empirical benefits of parallelization. We conclude that our parallel implementation performs best in almost all scenarios.

**Keywords:** Variable Neighborhood Search · Scatter Search · Nurikabe · Parallel algorithms

## 1 Introduction

Nurikabe which is a Japanese pencil game like Sudoku is played on an $n$ by $m$ grid of squares, some of which initially contain numbers. The goal of Nurikabe is to create a board that does not violate any of the following rules:

1. Every numbered cell must occupy a white region (an 'island') formed of contiguous white squares. The island must be sized to the number contained in the cell - for instance, an island that contains the integer '3' must be of size three. A violation of this rule can be seen in Fig. 2.
2. Black cells must form a contiguous wall around the individual islands. In other words, no black cell or set of black cells can be 'disconnected' from any other set of black cells. A violation of this rule can be seen in Fig. 3.
3. The 'wall' of black squares cannot, at any point, form a 2 by 2 block. A violation of this rule can be seen in Fig. 4.

The board shown in Fig. 1 is a valid Nurikabe solution.

Nurikabe is compared to Sudoku, a game on which extensive research has been conducted [1,2]. Comparatively, very little known is known about metaheuristic approaches to solving Nurikabe. Inspired by the results of [3], we propose an alternative hybrid algorithm that can be used to solve Nurikabe puzzles.

© Springer Nature Switzerland AG 2021
N. Mladenovic et al. (Eds.): ICVNS 2021, LNCS 12559, pp. 99–110, 2021.
https://doi.org/10.1007/978-3-030-69625-2_8

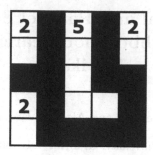

**Fig. 1.** A valid Nurikabe solution

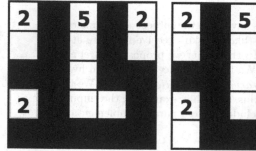

**Fig. 2.** An example of a violation of rule 1

**Fig. 3.** An example of a violation of rule 2

**Fig. 4.** An example of a violation of rule 3

Nurikabe is NP-complete as proved by Holzer et al. [4]. Their work caused researchers to shift focus from exact algorithms to metaheuristic and approximation algorithms. There exists, however, some work on direct solutions to Nurikabe [5]. Answer set programming has been used to create solutions to Nurikabe - however, due to Nurikabe's NP-complete status, such direct approaches may take longer time while solving the larger problem sizes.

In a recent paper, Amos et al. applied an ant colony metaheuristic algorithm to Nurikabe [3]. Rather than beginning with an all white board and adding black cells to create white islands - the method employed by the aforementioned direct solver - their algorithm begins by setting all non integer cells as walls and creates islands by sending out 'ants' from each integer cell. Their algorithm took less time than the their proposed logic based solver on smaller puzzles. Furthermore, the ant colony algorithm takes into account the game rules stated above, rather than attempting to generate a violation-free board from scratch using a brute logic approach.

We propose a hybrid metaheuristic algorithm for solving Nurikabe. Our algorithm combines Scatter Search [6] and Variable Neighborhood Search (VNS) [7]. At a high level, our algorithm generates a set of diverse solutions, selects the fittest subset of solutions, and searches the space around those solutions to find a correct, solved puzzle using a VNS.

We then parallelize our hybrid algorithm (Scattered-VNS), both allowing us to work with larger populations, and to cover more of the solution space with our neighborhood search.

## 2 Scattered-VNS for Nurikabe

Our algorithm can, roughly, be subdivided into two sub processes: board construction and neighborhood search.

### 2.1 Solution Representation

We encode each Nurikabe board using a $n$ by $m$ array of integers. We represent black squares using $-1$, white squares using $0$, and integer squares using the corresponding positive integer.

### 2.2 Fitness Function

We use the following formula to determine a given board's fitness:

$$(w_1 * D) + (w_2 * W) + (w_3 * B)$$

$D$, $W$, and $B$ are variables that, respectively, correspond to the following Nurikabe rule violations:

1. D - The sum total of the following formula for each island: |Island Integer − Island Size|, where 'island integer' denotes the integer found in the integer cell for a given island
2. W - The number of disjoint wall (black block) fragments
3. B - The number of $2 \times 2$ black blocks present on the board

Further, $w_1$ through $w_3$ determine how we weight each of these violations. These weights help us 'rank' how severe each violation is, helping us to answer questions like: if we fix a $2 \times 2$ block by increasing an island's size, causing said island to become larger than it ought to be, is our new board better or worse than our old one? We determined experimentally that the following weights allow us to solve the most puzzles: $w_1 = 1$, $w_2 = 2$, $w_3 = 5$.

### 2.3 Initial Solution: Board Construction

Our construction algorithm is based on *Scatter Search*. Scatter Search was first proposed by Fred Glover in 1977 [8] and it has been applied by researchers many times to solve various combinatorial optimization problems.

Scatter search generates a set of candidate solutions, selects the most fit subset of those solutions, and attempts to use that fit subset (using a crossover algorithm) to generate new, fitter candidate solutions [6]. Therefore, a Scatter Search can be explained in three steps:

1. Initial candidate generation
2. Candidate subset selection
3. Candidate crossover generation

Our construction algorithm employs (1) and (2), but does not use a crossover function. We first place the integer cells on the board, and initialize every other cell as a black square. We then create a pool of *wsq* white squares, where *wsq* is equal to the number of white squares that are present in a valid solution - in other words, boards are always initialized with the correct number of white cells.

A weighted random distribution is used to distribute white cells among existing islands such that no two islands overlap. This 'no-overlap' criteria will never be violated - all islands will always remain distinct. The weighted random distribution is more likely to add cells to islands with larger integers - in other words, an island containing 6 is twice as likely to receive a white square as an island that contains 3. This process ensures that islands are generated semi-randomly, but also that islands that should be larger end up with more white squares than those that should be smaller. We then repeat this process, creating a diverse (due to the random nature of our generation algorithm) set of population members. After generation, we select a top percentage of the boards by fitness, and use those as our population pool for the subsequent neighborhood search.

## 2.4 Neighborhood Structures

We propose four neighborhood structures. All neighborhoods must follow our 'no overlap' rule: white islands cannot overlap. So, any neighborhood operation in which a white cell is selected and removed does not split an island, and any neighborhood operation in which a black cell is removed or added neither splits a white island, nor merges two white islands.

**Surplus Island Swap.** This neighborhood is intended to help decrease the number of violation on island sizes. We first check which islands have too many white cells, adding all these islands to a set $L$. We then create a set of islands that are too small (call it $S$). We select an island $I_L$ from $L$ and an island $I_S$ from $S$. We then remove a white cell from $I_L$ without splitting it into two islands, and add a white cell to $I_S$ without creating an overlap with another island. This procedure decreases the size of islands that are too large and increases the size of islands that are too small. An illustration of this process can be seen in Fig. 5.

(a) Before surplus swap operation

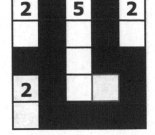
(b) After surplus swap operation

**Fig. 5.** Surplus swap operation

**Unify Wall.** This neighborhood is intended to help decrease the number of isolated black cells (or 'wall fragments'). We first make a list of wall fragments, then, using a selection weighted towards smaller fragments, select a random fragment. We then swap a random black square from the selected fragment with a random white square abutting another wall fragment. An illustration of this process can be seen in Fig. 6.

This, over time, unifies the black blocks into a single monolithic, contiguous structure. Since the selection process is weighted towards choosing smaller fragments, we are more likely to remove blocks from smaller fragments and add them to larger fragments.

(a) Before unify wall operation

(b) After unify wall operation

**Fig. 6.** Unify wall operation

**Break up 2 × 2 Black Blocks.** This neighborhood, like unify wall, selects a random block that both makes up a 2 × 2 block and abuts an island. We then swap that black block with a white block from an island, unless said swap would create another 2 × 2 square. An illustration of this process can be seen in Fig. 7.

(a) Before break up operation

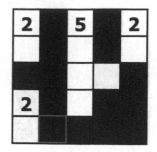
(b) After break up operation

**Fig. 7.** Break up $2 \times 2$ operation

**Regenerate Island.** This neighborhood is intended to help move our search out of 'ruts' in which none of the above neighborhoods can make moves without violating the overlap constraint. An illustration of this process can be seen in Fig. 8. We select a random island, set its size to 0, and then regenerate it. This generates a new island equal in size to the original, but with a different shape, thereby giving other neighborhoods new search options.

(a) Before the regeneration operation

(b) During the regeneration operation

(c) After regenerating the island

**Fig. 8.** Regeneration operation

### 2.5    Scattered-VNS

We combine our Scatter Search and VNS as shown in Algorithm 1. First four lines of the algorithm generate many boards using Scatter Search principle. For each board provided by our scatter search generation and selection method, our VNS iterates a certain number of times. For each VNS iteration, we run each neighborhood search once, attempting to find the first best improvement for that neighborhood. If we can find an improvement over the current global best (call this improved board $B_i$), we do two things. We first update the current and best boards, terminating if the best board's fitness value reaches 0, indicating that it has been solved. We then iterate through every other board (call a given board $B_o$) that hasn't been searched yet. During this iteration there is a small chance that $B_o$ will be replaced with $B_i$.

---

**Algorithm 1.** Scattered Variable Neighborhood Search

---

**procedure** SCATTEREDVNS(board, popSize, fitPercent, probReplace, LSIters)
    $popSize \leftarrow max(\text{puzzle x-dimension}, \text{puzzle y-dimension})^3$
    startingBoards[] $\leftarrow$ Generate(popSize, board)
    $\forall$ startingBoard in startingBoards, calculateFitness(startingBoard)
    topBoards[] $\leftarrow$ most fit fitPercent% of boards in startingBoards
    globalBest $\leftarrow$ topBoards[0]
    **for** currBoard in topBoards **do**
        **for** LSIters many iterations **do**
            currBoard $\leftarrow$ SurplusIsland(currBoard)
            currBoard $\leftarrow$ UnifyWall(currBoard)
            currBoard $\leftarrow$ BreakUp2x2(currBoard)
            currBoard $\leftarrow$ RegenerateIsland(currBoard)
            **if** Fitness(currBoard) == 0 **then**
                globalBest $\leftarrow$ currBoard
                Break
            **if** Fitness(currBoard) < Fitness(globalBest) **then**
                globalBest $\leftarrow$ currBoard
                **for** Board in topBoards **do**
                    Replace(Board, globalBest, probReplace)
    Return globalBest

---

## 3 Parallel Scattered-VNS for Nurikabe

We propose a parallel implementation of Scattered-VNS to solve Nurikabe. We employ the same solution representation and fitness function as Algorithm 1. While we employ the same neighborhood structures, we do remove the 'island replace' chance, and thus *do not* modify our population at all after the initial generation process.

Our sequential Scattered-VNS algorithm is parallelized on two levels. First, our root process (S1) generates a board population using the process discussed in Sect. 2.3. It then divides board population into M many subsets and distributes the sets among M many processes $(P_1, P_2...P_M)$. In our case, M equals 8. Then, in each of the processes, each of the K neighborhoods is run $(N1, N2...N_k)$ - in our case, K equals 4, as our neighborhoods are identical to those in Sect. 2.4. After each neighborhood finishes its search, it returns the best (fittest) board found, determining fitness using the previously discussed fitness function from Sect. 2.2. After all neighborhoods return, the parent process then takes the fittest of these boards, and returns it to the root process (S1). After all $M$ processes return, S1 then accepts the fittest board.

This parallelization strategy benefits from the lack of communication between population members in our sequential Scattered VNS algorithm. Since, in the parallel implementation, each subset of our population is independent of all other population subsets, we can avoid traditional parallelization slowdowns like locks and semaphores, and avoiding common threading bugs like race conditions. The same is true of each neighborhood thread. Conversely, our main source of slowdown comes from waiting for processes to terminate. We visualize this process in Fig. 9.

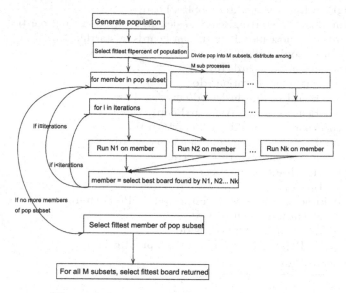

**Fig. 9.** A flowchart of Parallel Scattered-VNS

# 4   Experimental Results

## 4.1   Dataset

We use boards from Janko, a puzzle repository, as our algorithm's test dataset [9]. From this repository, we selected a set of 38 puzzles ranging from 9 to 255 cells. Our sequential Scattered-VNS and VNS algorithms were only tested on the first 20 boards.

## 4.2   Parameter Setting

Both algorithms, as seen above, takes the following parameters:

1. fitPercent - This parameter determines what percentage of the initialized boards are selected for our VNS
2. Iterations - This parameter determines how many neighborhood searches we run before moving on to the next candidate board
3. ProbReplace - This parameter determines how likely we are to replace a given board with a newly discovered global best - for our parallel implementation, this is set to 0.

Experimentally, we determined that our sequential algorithms performs best with a fitPercent value of 0.2, an iterations value of 100, and a probReplace value of 0.05. For our parallel algorithm, we set probReplace to 0 and (due to the time savings) increase the number of iterations to 250.

We ran and tested our algorithm using the same criteria as in [3], running our algorithm 30 times. If any of the 30 runs found a solution, the puzzle was marked as solved. Both algorithms were implemented using python 3.7[1], and tested on a machine running Ubuntu 18.04 using an Intel core i-7 8700k.

### 4.3 Results

Using the above criteria, we tested Sequential VNS, Sequential Scattered-VNS and Parallel Scattered-VNS on 20 boards. Sequential VNS uses the same neighborhoods, but does not use any of the scatter search elements (i.e. the neighborhoods are only run on one initially generated board). Sequential algorithms run 100 iterations. Results are listed in Table 1. The 'Cells' column keeps track of how many cells a given puzzle has - this can be used as a proxy for puzzle difficulty. The 'Hit %' column tracks robustness by tracking what fraction of attempts found a solution to the puzzle. Finally, 'Sol. Time' measures the number of seconds needed to solve a given puzzle. Unsolved boards, or boards that timed out[2], are denoted by 'unsol'. Because our Sequential VNS failed to solve the overwhelming majority of problems it was given, we have included a "best fitness" column. This column contains the fitness value of the best board the algorithm found, serving as a proxy for how "close" the algorithm got to a valid solution; fitness is calculated using the fitness function from Sect. 2.2.

Our Sequential Scattered-VNS was able to solve every problem. However, [3]'s ACO implementation was only allowed to run for one minute before a run was counted as a failure. Were our algorithm subjected to the same constraint, it would have failed to solve several of the above puzzles. Our Sequential Scattered-VNS algorithm is, per the provided hit percentage, quite robust. Every hit percentage except one is 100, meaning that every attempt found a valid solution for the puzzle. Our algorithm struggled with one puzzle in particular (19), implying that there are some edge cases, or particular kinds of puzzles, it is not well suited to solving. As can be seen in Figs. 10 and 11, our Scattered-VNS algorithm improves solution fitness very quickly, and does not get stuck in local optima.

Our Sequential VNS couldn't solve most of the puzzles it was given, and for the puzzles it could solve, had a very low hit percentage. However, for those puzzles that it did solve our Sequential VNS outperformed our Scattered-VNS

---

[1] Due to Python's global interpreter lock, which limits the number of actual threads running at a given time to 1, we used processes instead of threads in our parallel implementation. Since processes are consume more memory than threads, and data had to be duplicated across processes, this may have caused some slowdown due to high memory, and thus high swap, usage.

[2] More than five minutes elapsed before a solution was found.

**Table 1.** Results of Sequential VNS, Sequential Scattered-VNS and Parallel Scattered-VNS

| Board data | | Sequential VNS | | | Sequential Scattered-VNS | | Parallel Scattered-VNS | |
|---|---|---|---|---|---|---|---|---|
| ID | Cells | Sol. Time | Hit % | Best Fitness | Sol. Time | Hit % | Sol. Time | Hit % |
| 0 | 9 | 0.27 | 6 | 0 | 1.12 | 100 | 3.76 | 100 |
| 1 | 16 | 0.29 | 80 | 0 | 1.23 | 100 | 1.25 | 100 |
| 2 | 25 | 0.31 | 20 | 0 | 3.22 | 100 | 6.08 | 100 |
| 3 | 25 | unsol | 0 | 7 | 5.16 | 100 | 1.34 | 100 |
| 4 | 25 | unsol | 0 | 3 | 2.29 | 100 | 1.78 | 100 |
| 5 | 25 | 0.44 | 23 | 0 | 4.46 | 100 | 3.83 | 100 |
| 6 | 25 | unsol | 0 | 4 | 2.62 | 100 | 1.58 | 100 |
| 7 | 25 | unsol | 0 | 6 | 4.26 | 100 | 2.33 | 100 |
| 8 | 25 | 0.91 | 13 | 5 | 3.17 | 100 | 1.07 | 100 |
| 9 | 36 | unsol | 0 | 11 | 6.49 | 100 | 3.01 | 100 |
| 10 | 36 | unsol | 0 | 6 | 7.13 | 100 | 6.78 | 100 |
| 11 | 36 | unsol | 0 | 12 | 17.49 | 100 | 7.84 | 100 |
| 12 | 36 | unsol | 0 | 5 | 22.73 | 100 | 4.93 | 100 |
| 13 | 36 | unsol | 0 | 7.5 | 9.96 | 100 | 2.66 | 100 |
| 14 | 36 | unsol | 0 | 5 | 12.44 | 100 | 6.16 | 100 |
| 15 | 36 | unsol | 0 | 5 | 14.72 | 100 | 5.92 | 100 |
| 16 | 36 | unsol | 0 | 12 | 25.55 | 100 | 15.03 | 100 |
| 17 | 36 | unsol | 0 | 7 | 31.61 | 100 | 9.27 | 100 |
| 18 | 36 | unsol | 0 | 11 | 16.38 | 100 | 8.49 | 100 |
| 19 | 36 | unsol | 0 | 7 | 38.55 | 3 | 22.72 | 10 |

algorithm in terms of speed. This indicates that the hybrid nature of our algorithm dramatically improves its robustness at the cost of increased run-time.

Table 1 also demonstrates that (per the provided hit percentages) our parallel Scattered-VNS both performs better and runs faster than the sequential version. While our parallel implementation didn't solve puzzle 37 as shown in Table 2, it reached a final fitness level of 14, a large improvement from the initial board's fitness level of 164. If we take [3]'s one minute limit into account, the Sequential Scattered-VNS solved 23/33 Nurikabe boards. The parallel version, over the same 32 boards, solved 31/33, a substantive increase.

Interestingly, the efficiency benefits of our parallel implementation only begin to manifest for larger boards - in smaller boards the parallel algorithm sometimes performs worse (as in the case of board 2, where it takes twice as long). The most likely explanation for this is that the constant-time overhead of creating 32 processes is quite large, and that with a smaller board, the constant time overhead of creating a thread overshadows any benefits threading may have.

Our algorithm also possibly performed better due to the removal of the probReplace variable and the increase in the number of iterations. By removing probReplace, we ensured that the population remained diverse over time, and by increasing the number of iterations we increased the portion of the solution space our neighborhood search was able to cover. Further, these results suggest that increasing the number of iterations does not result in this algorithm getting

**Table 2.** Results of sequential and parallel scattered-VNS on larger Nurikabe boards

| Board data | | Sequential scattered-VNS | Parallel scattered-VNS | |
|---|---|---|---|---|
| ID | Cells | Sol. Time | Sol. Time | Hit percentage |
| 20 | 36 | 52.22 | 33.25 | 100 |
| 21 | 36 | 193.29 | 25.34 | 100 |
| 22 | 36 | 102.82 | 19.57 | 100 |
| 23 | 49 | 34.15 | 33.29 | 100 |
| 24 | 49 | 71.22 | 32.45 | 100 |
| 25 | 49 | 23.15 | 43.66 | 100 |
| 26 | 49 | 74.15 | 58.29 | 100 |
| 27 | 49 | 91.15 | 37.45 | 100 |
| 28 | 64 | 246.15 | 53.66 | 20 |
| 29 | 64 | 102.21 | 41.25 | 76 |
| 30 | 81 | 132.32 | 52.13 | 43 |
| 31 | 100 | 201.11 | 71.87 | 50 |
| 32 | 100 | 231.74 | 68.29 | 26 |
| 33 | 121 | – | 81.38 | 13 |
| 34 | 144 | – | 99.55 | 16 |
| 35 | 169 | – | 147.82 | 3 |
| 36 | 196 | – | 162.05 | 6 |
| 37 | 225 | – | 204.76 | 0 |

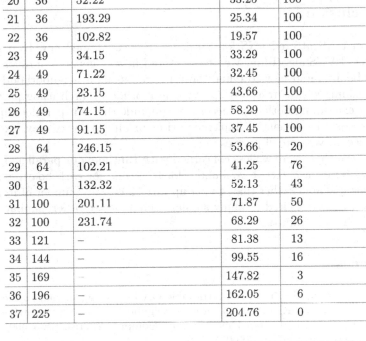

**Fig. 10.** Convergence graph of solution fitness over time for scattered VNS solution to puzzle 9

**Fig. 11.** Convergence graph of solution fitness over time for scattered VNS solution to puzzle 14

stuck in local optima, suggesting that further increasing the number of iterations could result in even better performance.

Further, we argue that our parallel algorithm would perform even better on newer, higher core hardware. Since the 8700k only has 12 logical processors, all 32 processes cannot run simultaneously. Therefore, on a higher core processor our parallel algorithm should, in theory, perform even better compared to the non-parallel version. Since processor core counts are increasing every year while single-core clock speeds aren't[3], the parallel version of our algorithm is both better today *and* is far more future-proofed.

## 5   Conclusion

We proposed a Scattered-VNS for Nurikabe, then examined how to parallelize our Scattered-VNS algorithm. We first explored how to generate a diverse set of Nurikabe boards, and how to refine those boards using a neighborhood search. We then examined how to speed up our neighborhood search using parallelization. After examining both approaches, we conclude that our parallel implementation exceeds the sequential implementation in almost every way, except for performance on very small puzzles.

We believe that, were we less concerned with runtime, our parallel algorithm would perform even better if we found a local optima for each board generated by our scatter search rather than giving up after a certain number of iterations.

Further, we are excited to apply similar techniques to other, similar puzzles like Ken-Ken and Kakuro.

## References

1. Das, K.N., Bhatia, S., Puri, S.: A retrievable GA for solving sudoku puzzles (2011)
2. Sevkli, A.Z., Hamza, K.A.: General variable neighborhood search for solving sudoku puzzles: unfiltered and filtered models. Soft Comput. **23**(15), 6585–6601 (2019). https://doi.org/10.1007/s00500-018-3307-6
3. Amos, M., Crossley, M., Loyd, H.: Solving nurikabe with ant colony optimization (extended version) *, June 2019. https://doi.org/10.13140/RG.2.2.23813.19685
4. Holzer, M., Klein, A., Kutrib, M.: On the NP-completeness of the nurikabe pencil puzzle and variants thereof (2008)
5. Cayli, M., Kavlak, E., Kaynar, H., Türe, F., Erdem, E.: Solving challenging grid puzzles with answer set programming. In: Proceedings of ASP, pp. 175–190, December 2007
6. Glover, F., Laguna, M., Marti, R.: Fundamentals of scatter search and path relinking. Control Cybern. **29**, 653–684 (2000)
7. Moreno Pérez, J.A., Hansen, P., Mladenovi'c, N.: Variable neighbourhood search: methods and applications. Ann. Oper. Res. **175**(01), 367–407 (2010)
8. Glover, F.: Heuristics for integer programming using surrogate constraints. Decis. Sci. **8**, 156–166 (1977). https://doi.org/10.1111/j.1540-5915.1977.tb01074.x
9. Janko, O.: Home. https://www.janko.at/Raetsel/Nurikabe/

---

[3] As of the time of writing.

# A Hybrid VNS for the Multi-product Maritime Inventory Routing Problem

Nathalie Sanghikian⑩, Rafael Martinelli⁽⊠⁾⑩, and Victor Abu-Marrul⑩

Industrial Engineering Department, Pontifical Catholic University of Rio de Janeiro (PUC-Rio),
Rio de Janeiro, Brazil
nathaliesanghikian@aluno.puc-rio.br, martinelli@puc-rio.br,
victor.cunha@tecgraf.puc-rio.br

**Abstract.** In a growth scenario of the world economy, it is essential to increase the integration between the different actors in the companies' supply chain, reducing operational costs, and improving efficiency. Ship routing is a crucial part of this integration regarding global maritime commerce. In this work, we present a hybrid VNS metaheuristic to tackle a real Maritime Inventory Routing Problem (MIRP) in a company that explores oil and gas in the Brazilian offshore basin. In the methodology proposed, a linear mathematical model is embedded in the local search procedure to minimize inventory costs. The approach, validated within realistic data, provides low and not regular inventory violations. When compared with a previously developed method, it presents an improved performance, with reduced costs and computational time.

**Keywords:** Maritime Inventory Routing Problem · Oil and gas industry · Variable Neighborhood Search

## 1 Introduction

In offshore oil and gas exploration, operating companies often have to manage products' transportation within their fleet, respecting inventory levels at ports and vessels in an integrated manner. The combination of these elements takes into account the ship's routing and scheduling with the inventory management, known in the literature as Maritime Inventory Routing Problem (MIRP) [3]. In these problems, the routing and scheduling steps define which ports will be visited by each vessel, in which sequence and time. Meanwhile, the inventory management step determines each port and vessel's inventory levels at each instant of time within a given planning horizon [3,5].

This work deals with a MIRP with multiple products applied to a Brazilian offshore oil and gas company to minimize operational and routing costs, respecting the ship's capacities and inventory levels in the ports. The company is responsible for the entire production process, from oil exploration, refining, and transportation. The company's planners are responsible for scheduling their available fleet, focusing on meeting specific demands and their respective delivery dates. da Costa [5] developed a combination of relax-and-fix and fix-and-optimize heuristics, testing the approach in a set of ten real-life instances. We extend their work by introducing a hybrid Variable Neighborhood

© Springer Nature Switzerland AG 2021
N. Mladenovic et al. (Eds.): ICVNS 2021, LNCS 12559, pp. 111–122, 2021.
https://doi.org/10.1007/978-3-030-69625-2_9

Search (VNS) approach to solve the problem. The method combines the VNS structure to optimize the routes with an embedded linear mathematical formulation used to optimize the inventory levels. Moreover, the proposed model contemplates the possibility of transforming products, which represents adaptability for the model to choose which product to use to serve a customer with flexible quality demand [5]. The main contributions of our work are: (1) Propose a new hybrid approach that combines the VNS structure with a linear mathematical formulation to deal with a realistic MIRP in reasonable computational times; (2) Address a real-life multi-product MIRP, considering ten instances with real data from a Brazilian offshore oil and gas company.

The outline of the paper is organized as follows. Section 2 presents a brief review of works applying heuristics to solve different realistic and theoretical variations of MIRP. Section 3 describes the real multi-product MIRP problem considered in this paper. In Sect. 4, the proposed hybrid VNS is presented, including the linear mathematical formulation used to optimize the inventory levels. In Sect. 5, computational experiments are conducted as discussed, considering a set of real instances provided by the studied company. Finally, conclusions and future remarks are given in Sect. 6.

## 2 Literature Review

The basic MIRP and some of its extensions are described by Christiansen and Fagerholt [3]. The basic case considers only one product, known storage capacities in the ports, and constant production and demand rates. The main goal is to develop routes and schedules for a fleet of ships, minimizing transportation and inventory costs, and meeting the demand within a given planning horizon. Real-life problems are more complex, with other aspects, such as consumer or central supplier, stock constraints for distinct subsets of ports, variable production or demand rates, several products, charters, and others.

Several authors have applied heuristics and hybrid methods to solve different MIRPs in the literature, in theoretical and real-life contexts. Dauzère-Pérès et al. [6] designed a decision support system using a memetic algorithm to deal with a problem related to a calcium carbonate paste supplier. Christiansen et al. [4] dealt with a real-life problem in a cement industry using a multi-start constructive procedure guided by a genetic algorithm. They focused on maximizing a multi-criteria objective function, defining a weight for each criterion. Siswanto et al. [14] addressed a ship routing and scheduling problem with many technical and physical constraints and non-dedicated compartments. The authors developed a hybrid approach with a Mixed Integer Linear Programming (MILP) formulation, using heuristics to solve sub-problems of route selection, vessel selection, loading, and unloading quantities. Song and Furman [15] also used a hybrid approach combining mathematical formulation with heuristics to solve sub-problems and improve a set of given solutions for a MIRP. Uggen et al. [17] has developed a heuristic approach based on relax-and-fix and fix-and-optimize for a MIRP. Hemmati et al. [9] proposed a two-phase iterative hybrid metaheuristic called Hybrid Cargo Generating and Routing to solve a short distance inventory routing problem with multiple products. In the first phase, a simplified version of the problem is solved by a mathematical formulation. An Adaptive Large Neighborhood Search is applied to improve the solutions in the second phase. Diz et al. [7] developed relax-and-fix and fix-and-optimize heuristics to deal with a real problem related to the Brazilian offshore oil industry. Papageorgiou et al. [12] presented an extensive computational study, comparing variants of rolling horizon heuristics, K-opt heuristics, local branching, solution polishing, and hybrid approaches to solve a MIRP. Munguía et al. [11] developed a hybrid approach using MILP formulations to solve sub-problems iteratively. Bertazzi et al. [2] designed a three-phase matheuristic to solve a MIRP. Their approach grouped costumers by similarity (clustering phase), designing routes based on the clusters (routing construction phase), and searching for improvements on the solution using a binary linear mathematical formulation (optimization phase). da Costa [5] handled a realistic multi-product MIRP from a Brazilian oil and gas company. The objective was to develop a decision support tool to automate the company's ship scheduling process. The author applied a combination of relax-and-fix and fix-and-optimize heuristics to solve the problem, overcoming the company's current solutions, with an average execution time of three hours. We refer the reader to the work of Papageorgiou et al. [13] for a complete overview of the MIRP literature.

We extend the work of da Costa [5] by developing a hybrid VNS to deal with the same multi-product MIRP from a Brazilian oil and gas company, intending to develop improved solutions to the practical problem with less computational time. A new mathematical formulation is introduced, based on the one presented by Christiansen and Fagerholt [3], using continuous instead of discrete-time, allowing a vessel to stay for less than one day in each port.

## 3  Problem Description

Let $G = (\mathcal{N}, \mathcal{E})$ denote a graph where $\mathcal{N}$ is the set of *ports*, and $\mathcal{E}$ the set of edges connecting them. To meet a given demand, the company has a heterogeneous fleet of *ships*, defined in the set $\mathcal{V}$, transporting different oil *products*, defined in the set $\mathcal{P}$, between ports within a given planning horizon $H$. Each port $i \in \mathcal{N}$ has a product handling rate $R_i$, and a known production or demand $PD_{ip}$ for product $p \in \mathcal{P}_i$ (positive values represent production, while negative ones represent demands), where $\mathcal{P}_i \subseteq \mathcal{P}$ is the subset of products that are produced or demanded by port $i$. Moreover, inventory levels, at each port $i \in \mathcal{N}$ for each product $p \in \mathcal{P}_i$, must be within a given interval $[S_{ip}^{MN}, S_{ip}^{MX}]$ during the planning horizon. To meet the demand, the company has a heterogeneous fleet of ships, defined in the set $\mathcal{V}$, transporting the products between ports. The time taken by a ship $v$ to traverse an edge $(i, j) \in \mathcal{E}$ is given by $T_{ijv}$. Each ship $v \in \mathcal{V}$ has a capacity $K_v$ and it is only able to visit ports in the subset $\mathcal{N}_v \subseteq \mathcal{N}$. Draft constraints are also considered in the problem, taking into account the cargo on the ships and the characteristics of the ports. To accomplish the draft constraints, the total loading on-board each ship $v$ when visiting port $i$ must be within a given interval $[L_{iv}^{MN}, L_{iv}^{MX}]$. Ports and ships may have initial inventory levels, indicated by parameters $S_{0ip}$ and $L_{0vp}$, respectively.

As mentioned above, each port has a demand or production for a specific product $p \in \mathcal{P}_i$. However, some demands can be met by different products, as long as they have the necessary quality. Thus, we define $\mathcal{P}_p$ as the subsets of products allowed to meet a specific demand for product $p$. This transformation allows the model the possibility to choose which product to use to meet demand with flexible quality, aiming for the lowest inventory violation and cost. The transformation of one product into another is only allowed during a ship's loading operation in a port.

The objective is to define routes for the available ships, stipulating which ports to visit and the number of products to be loaded and unloaded at each visit, minimizing routing and operational costs. Each port can be visited several times by the same ship during the planning horizon. Route's costs relate to the total distance that each ship must travel in a given solution, where $C_{ij}^T$ is the traveling distance associated to edge $(i, j) \in \mathcal{E}$. Operational costs regard product handling and inventory holding costs at the ports. $C_{iv}^H$ is the handling cost of loading or unloading one product unit by ship $v$ at port $i$, while $C_i^S$ is the inventory holding cost at port $i$. Products delivered to ports might generate earnings that reduce operational costs. $E_{ip}^U$ is the amount that the company earns for each product $p \in \mathcal{P}$ unit delivered at port $i \in \mathcal{N}$.

To solve the problem, we developed a hybrid VNS approach, presented in the next section, where neighborhood procedures define the vessel's routes, while the total operational cost (considering inventory holding costs, cargo handling costs, delivery earnings, and violation penalties) is optimized by a linear mathematical formulation embedded in the local search. In real applications, it is challenging to find solutions for the problem that meets the limits of cargo on-board ships and draft and inventory limits at the ports. Based on this, we relax the draft and inventory limits constraints, penalizing each violation with a parameter $\mu$. The proposed method searches for solutions with minimum penalties, aiming to achieve feasible solutions if possible.

## 4 Hybrid VNS Approach

In this section, we detail our hybrid VNS approach. VNS is a metaheuristic based on a systematic change of neighborhoods to solve combinatorial optimization problems. It has been successfully applied to several logistic problems in the literature [8]. Our hybrid VNS applies perturbation movements and local searches iteratively, searching for neighbor solutions that improve the current solution. Each neighbor solution is defined by a single change in the routing structure of the current solution, triggering a change in the inventory values obtained by a linear mathematical formulation.

To represent the solution, we introduce the concept of a port call. The calls account for the number of port visits during the planning horizon regardless of the ship that performs it. Thus, if a port $i$ is visited twice by different vessels, two calls are performed in this port. Therefore, we define a set $\mathcal{M}_i$ of calls to sequence the visits in the ports. A specific solution for the MIRP can be defined as a permutation of pairs $(i, m) \in \mathcal{N}_v \times \mathcal{M}_i$, indicating the sequence of ports that each ship visits with its respective call.

Figure 1 shows an example of a solution following the proposed structure, with each element on the solution represented as a pair indicating a port and its call. Two ships and four ports are considered in the case given, with up to two calls performed at each port. Note that *Ship 1* and *Ship 2* visit the ports in the following order: 1–2–4–3, and 2–3–1, respectively. Regarding *Port 1*, its first call is performed by *Ship 2*, followed by a visit from *Ship 1*. Note that when solving the inventory part of the problem, *Visit* $(1, 2)$ of *Ship 1* will only be performed after the completion of *Ship 2* route, due to *Visit* $(1, 1)$.

| Ship 1 | (1, 2) | (2, 2) | (4, 1) | (3, 2) |
|--------|--------|--------|--------|--------|

| Ship 2 | (2, 1) | (3, 1) | (1, 1) |
|--------|--------|--------|--------|

**Fig. 1.** Solution representation considering an instance with two ships and four ports.

The Hybrid VNS proposed in this work, presented in Algorithm 1, needs two parameters to run: the total number of iterations ($\eta$) and the maximum number of neighborhoods used during the local search ($\ell_{max}$). We designed four different neighborhoods, using $\ell_{max}$ equal to four in every execution of the VNS. The proposed neighborhoods are as follows.

- Swap: exchanges two visits in the solution;
- Relocate: removes a visit in the solution, inserting in another position;
- Insert: inserts a new visit in a route;
- Delete: deletes a visit from a route.

The first two neighborhoods are considered in its intra-route (on the same route) and inter-route (between two different routes) versions. The Insert neighborhood always inserts a visit to the next possible call of the port, and the Delete neighborhood always deletes the visit for the last call of the port. Regardless of which neighborhood is selected, the solution cannot result in the same port being visited consecutively by the same ship.

The algorithm starts by building the set $\mathcal{L}$ of neighborhoods to be used during the local search step (Line 1). Then, it calls a procedure that builds an initial solution $s$ considering only each ship's initial position, calling the local search before returning the solution. Thus, we do not define initial routes for the ships, allowing the local search to build the solution iteratively while the mathematical formulation optimizes the inventory levels. The main loop is executed $\eta$ times (Lines 4–23) and consists of a perturbation procedure that moves the solution $s$ to a random neighbor solution $s'$ in neighborhood $k$ (Lines 5–6), followed by a Randomized Variable Neighborhood Descent (RVND) [16] local search applied on $s'$ (Lines 7–16). The RVND procedure starts by shuffling the set of neighborhoods $\mathcal{L}$ (Line 7) and running the local search following the defined random order of neighborhoods (Lines 8–16). For each neighborhood $\ell$, the mathematical formulation is executed to optimality (Line 10), and the algorithm checks whether the new neighbor solution $s''$ is better than $s'$, updating $s'$ accordingly, moving to the next neighborhood if true. After the RVND, the algorithm decides if the new solution $s'$ will be accepted by a pre-established criterion (Lines 17–22). This criterion is fulfilled if the cost of the new solution $s'$ is better than the cost of $s$. Also, it is possible to accept a worse solution by a probability factor [10]. Once accepted, $s$ is updated, and then it is verified if $s$ is better than the best solution obtained so far $s^*$ (Lines 19–21). The method returns the final solution $s^*$ in Line 24.

---

**Algorithm 1:** Hybrid VNS $(\eta, \ell_{max})$

---

1   $\mathcal{L} \leftarrow \{1, .. , \ell_{max}\}$
2   $s \leftarrow Construct()$
3   $s^* \leftarrow s$
4   **for** $\eta$ *iterations* **do**
5      $k \leftarrow Random(\mathcal{L})$
6      $s' \leftarrow Perturb(s, k)$
7      $shuffle(\mathcal{L})$
8      **for** $\ell \in \mathcal{L}$ **do**
9          **for** $s'' \in N_\ell(s')$ **do**
10              $s'' \leftarrow Formulation(s'')$
11              **if** $f(s'') < f(s')$ **then**
12                  $s' \leftarrow s''$
13                  *break*
14              **end**
15          **end**
16      **end**
17      **if** $Accept(s', s)$ **then**
18          $s \leftarrow s'$
19          **if** $f(s) < f(s^*)$ **then**
20              $s^* \leftarrow s$
21          **end**
22      **end**
23 **end**
24 **return** $s^*$

---

### 4.1 Mathematical Formulation

The proposed mathematical formulation is based on the one described by Christiansen and Fagerholt [3], but disregarding routing variables and including some particular constraints related to the real problem. Based on a given routing solution, we define $\mathcal{M}_i$ as the set of calls on port $i$, and $\mathcal{M}_{iv}$ as the set of calls on port $i$ performed by ship $v$. A parameter $M_i^F$ indicates the last call on port $i$. Moreover, the network flow part of the formulation works on an extended graph, considering nodes $(i, m) \in \mathcal{N}_v \times \mathcal{M}_{iv}$, and arcs $(i, m, j, n) \in \mathcal{A}_v \subseteq (\mathcal{N}_v \times \mathcal{M}_{iv}) \times (\mathcal{N}_v \times \mathcal{M}_{iv})$, for each ship $v$. Moreover, each ship $v \in \mathcal{V}$ starts at port $\mathcal{N}_v^0$ performing call $M_v^0$. Finally, let $V_{im}^B$ be the ship performing the call just before call $m$ on port $i$, i.e., $(m-1) \in \mathcal{M}_{iV_{im}^B}$, and $V_i^F$ be the ship performing the last call of port $i$, i.e., $M_i^F \in \mathcal{M}_{iV_i^F}$.

The amount loaded and unloaded by each ship $v \in \mathcal{V}$, on each port $i \in \mathcal{N}_v$, in call $m \in \mathcal{M}_{iv}$, of each product $p \in \mathcal{P}_i$, are given by variables $q_{ivmp}^L$ and $q_{ivmp}^U$, respectively. Variables $s_{imp}$ indicate the inventory level on port $i \in \mathcal{N}$ in call $m \in \mathcal{M}_i$ of each product $p \in \mathcal{P}_i$, while variables $l_{ivmp}$ compute the total load on-board ship $v \in \mathcal{V}$ when visiting port $i \in \mathcal{N}_i$ to perform call $m \in \mathcal{M}_{iv}$ of product $p \in \mathcal{P}$. As mentioned before, we use a penalty strategy to accept infeasible solutions. To achieve this, two slack variables $s_{imp}^{slack}, i \in \mathcal{N}, m \in \mathcal{M}_i, p \in \mathcal{P}_i$, and $l_{ivm}^{slack}, v \in \mathcal{V}, i \in \mathcal{N}_v, m \in \mathcal{M}_{iv}, p \in \mathcal{P}_i$ are included to allow inventory levels outside the limit ranges. To obtain the scheduling, variables $t_{im}$

indicate the time in which a call $m \in \mathcal{M}_i$ at port $i \in \mathcal{N}$ starts. Variables $r_{imvpp'}, v \in \mathcal{V}, i \in \mathcal{N}_v, m \in \mathcal{M}_{iv}, p \in \mathcal{P}_i, p' \in \mathcal{P}_p$ indicates the amount of product $p'$ replacing the demand of product $p$ during a loading operation in call $m$ of ship $v$ at port $i$. The formulation to minimize the operational costs including the objective function is as follows.

$$\min \sum_{i \in \mathcal{N}} \sum_{m \in \mathcal{M}_i} \sum_{p \in \mathcal{P}_i} C_i^S s_{imp} + \qquad \text{(inventory cost)}$$

$$\sum_{v \in \mathcal{V}} \sum_{i \in \mathcal{N}_v} \sum_{m \in \mathcal{M}_{iv}} \sum_{p \in \mathcal{P}_i} C_{iv}^H (q_{ivmp}^L + q_{ivmp}^U) - \qquad \text{(waiting and handling costs)}$$

$$\sum_{v \in \mathcal{V}} \sum_{i \in \mathcal{N}_v} \sum_{m \in \mathcal{M}_{iv}} \sum_{p \in \mathcal{P}_i} E_{ip}^U q_{ivmp}^U + \qquad \text{(delivery earnings)} \qquad (1)$$

$$\mu \left( \sum_{i \in \mathcal{N}} \sum_{m \in \mathcal{M}_i} \sum_{p \in \mathcal{P}_i} s_{imp}^{slack} + \sum_{v \in \mathcal{V}} \sum_{i \in \mathcal{N}_v} \sum_{m \in \mathcal{M}_{iv}} l_{ivm}^{slack} \right) \qquad \text{(infeasibility penalization)}$$

subject to

$$l_{jvnp} = l_{ivmp} + q_{jvnp}^L - q_{jvnp}^U - \sum_{p' \in \mathcal{P}_p} r_{jvnpp'} + \sum_{p' \in \mathcal{P}_p} r_{jvnp'p} \qquad \begin{array}{c} \forall v \in \mathcal{V}, (i, m, j, n) \in \mathcal{A}_v, \\ p \in \mathcal{P} \end{array} \qquad (2)$$

$$L_{iv}^{MN} - l_{ivm}^{slack} \leq \sum_{p \in \mathcal{P}} l_{ivmp} \leq L_{iv}^{MX} + l_{ivm}^{slack} \qquad \forall v \in \mathcal{V}, i \in \mathcal{N}_v, m \in \mathcal{M}_{iv} \qquad (3)$$

$$L_{jv}^{MN} - l_{jvn}^{slack} \leq \sum_{p \in \mathcal{P}} l_{ivmp} \leq L_{jv}^{MX} + l_{jvn}^{slack} \qquad \forall v \in \mathcal{V}, (i, m, j, n) \in \mathcal{A}_v \qquad (4)$$

$$q_{ivmp}^L \leq l_{ivmp} \qquad \begin{array}{c} \forall v \in \mathcal{V}, i \in \mathcal{N}_v, m \in \mathcal{M}_{iv}, \\ p \in \mathcal{P}_i \end{array} \qquad (5)$$

$$\sum_{p \in \mathcal{P}} l_{ivmp} \leq K_v - \sum_{p \in \mathcal{P}_i} q_{ivmp}^U \qquad \forall v \in \mathcal{V}, i \in \mathcal{N}_v, m \in \mathcal{M}_{iv} \qquad (6)$$

$$t_{jn} \geq t_{im} + T_{ijv} + \sum_{p \in \mathcal{P}} \frac{(q_{ivmp}^L + q_{ivmp}^U)}{R_i} \qquad \forall v \in \mathcal{V}, (i, m, j, n) \in \mathcal{A}_v \qquad (7)$$

$$t_{im} \geq t_{i(m-1)} + \sum_{p \in \mathcal{P}} \frac{\left( q_{iV_{im}^B (m-1)p}^L + q_{iV_{im}^B (m-1)p}^U \right)}{R_i} \qquad \forall i \in \mathcal{N}, m \in \mathcal{M}_i \setminus \{1\} \qquad (8)$$

$$s_{i1p} = S_{0ip} + (PD_{ip} t_{i1}) \qquad \forall i \in \mathcal{N}, p \in \mathcal{P}_i \qquad (9)$$

$$s_{imp} = s_{i(m-1)p} + (q_{iV_{im}^B (m-1)p}^U - q_{iV_{im}^B (m-1)p}^L) + PD_{ip}(t_{im} - t_{i(m-1)}) \qquad \begin{array}{c} \forall i \in \mathcal{N}, m \in \mathcal{M}_i \setminus \{1\}, \\ p \in \mathcal{P}_i \end{array} \qquad (10)$$

$$S_{ip}^{MN} - s_{imp}^{slack} \leq s_{imp} \leq S_{ip}^{MX} + s_{imp}^{slack} \qquad \forall i \in \mathcal{N}, m \in \mathcal{M}_i, p \in \mathcal{P}_i \qquad (11)$$

$$s_{iM_i^F p} - (q_{iV_i^F M_i^F p}^L - q_{iV_i^F M_i^F p}^U) + PD_{ip}(H - t_{iM_i^F}) \leq S_{ip}^{MX} + s_{iM_i^F p}^{slack} \qquad \forall i \in \mathcal{N}, M_i^F \neq 0, p \in \mathcal{P}_i \qquad (12)$$

$$s_{iM_i^F p} - (q_{iV_i^F M_i^F p}^L - q_{iV_i^F M_i^F p}^U) + PD_{ip}(H - t_{iM_i^F}) \geq S_{ip}^{MN} - s_{iM_i^F p}^{slack} \qquad \forall i \in \mathcal{N}, M_i^F \neq 0, p \in \mathcal{P}_i \qquad (13)$$

$$t_{im} \leq H \qquad \forall i \in \mathcal{N}, m \in \mathcal{M}_i \qquad (14)$$

$$l_{\mathcal{N}_v^0 v M_v^0 p} = L_{0vp} \qquad \forall v \in \mathcal{V}, p \in \mathcal{P}_i \qquad (15)$$

$$q_{ivmp}^L, q_{ivmp}^U, l_{ivmp} \geq 0 \qquad \begin{array}{c} \forall v \in \mathcal{V}, i \in \mathcal{N}_v, m \in \mathcal{M}_{iv}, \\ p \in \mathcal{P}_i \end{array} \qquad (16)$$

$$l_{ivmp}^{slack} \geq 0 \qquad\qquad\qquad\qquad \forall v \in \mathcal{V}, i \in N_v, m \in M_{iv}, \qquad (17)$$

$$s_{imp}, s_{imp}^{slack} \geq 0 \qquad\qquad\qquad\qquad \forall i \in N, m \in M_i, p \in \mathcal{P}_i \qquad (18)$$

$$t_{im} \geq 0 \qquad\qquad\qquad\qquad\qquad \forall i \in N, m \in M_i \qquad\qquad (19)$$

$$r_{imvpp'} \geq 0 \qquad\qquad\qquad\qquad \begin{aligned} &\forall v \in \mathcal{V}, i \in N, m \in M_i, \\ &p \in \mathcal{P}_i, p' \in \mathcal{P}_p \end{aligned} \qquad (20)$$

As mentioned before, the objective function (1) minimizes the operational costs, considering holding and inventory costs at the ports, unloading earnings and infeasibility penalties. Constraints (2) compute the cargo on-board the ship at each port call. Constraints (3) and (4) limit the total cargo on-board the ship to be within interval $[L_{iv}^{MN}, L_{iv}^{MX}]$, with the slack variable $l_{ivm}^{slack}$ used to relax the constraints, allowing violations in the limits. Constraints (5) and (6) limit the cargo on-board the ship to respect the ship capacity and the amount to load and unload. Constraints (7) and (8) compute the starting time of a port call. Constraints (9) and (10) compute the inventory level at the ports at each call. Constraints (11)–(13) force the inventory level to respect limits $[S_{ip}^{MN}, S_{ip}^{MX}]$, using the slack variable $s_{imp}^{slack}$ to relax these constraints, allowing limit violations. Constraints (14) limit the time to start visits to be within the planning horizon. Constraints (15) set the initial load on each ship. Constraints (16)–(20) specify the variables' domains.

## 5    Computational Experiments

This section presents the results of the experiments conducted on a set of ten real-life MIRP instances from a Brazilian oil and gas company. All experiments were performed on a computer with an Intel i7-8700K CPU of 3.70 GHz and 64 GB of RAM running Linux. The Hybrid VNS and the mathematical formulation were coded using Julia language v1.0.5, and the model was solved by CPLEX 12.8 solver, running in a single thread. We ran each instance ten times, limiting the Hybrid VNS execution to 250 iterations.

### 5.1    Instances Details

In the set of ten realistic MIRP instances considered in this work, ports are split into national (located on the Brazilian coast) or international ones. National ports are producers and consumers, while international ports are exclusively consumers. The demand for international ports is not mandatory but generate earnings for the company. Each instance is composed of 18 ports (15 national and three international ports), a fleet size ranging between 12 and 15 ships, nine different oil products, and a planning horizon of about 60 days. Another critical aspect of the instances is that, in general, it is hard to find feasible solutions for them. In many cases, the company needs to find alternative ways to dispose of excess production or meet incomplete demands, increasing the operational costs.

## 5.2 General Results

Aiming to compare the obtained results with the Hybrid VNS method, we took the final routes generated by da Costa [5] and calculated their solution values using our approach. The operational costs not considering the penalties are shown in Table 1. The table shows the name of each instance in the first column, followed by their Best-Know Solutions (BKS), considering routing and operational costs, disregarding penalization costs. The hybrid VNS results are presented in terms of the minimum cost (*min*), average cost (*avg*), maximum cost (*max*), standard deviation (*sd*), and average computational time ($\overline{time}$), for ten runs on each instance. The results indicate a significant difference in the costs of our solutions and the ones obtained by da Costa [5].

One can note that new BKSs are provided for all instances, as the VNS improves the solutions obtained by da Costa [5], even if we consider only the worst-case solutions (*max*). Another important aspect concerns the low values for standard deviations. It shows that the method performs well in realistic instances, providing solutions with low variability in practice, regardless of the chosen seed. When it comes the computational time, the results are also good. The worst-case (instance 20171011) had an average computational time of approximately 26 min.

## 5.3 Violations Analysis

In this section, we compare the violations found by each approach, regarding inventory and draft limits in ports. In Fig. 2, we analyse the distributions of the violations for the minimum inventory level at each port ($S_{ip}^{MN}$), shown in Fig. 2(a), and for the maximum inventory limit at each port ($S_{ip}^{MX}$), shown in Fig. 2(b). We present the violations in a $\log_{10}$ scale for better visualization.

Note that, in both cases, the VNS performs considerably better than the method proposed by da Costa [5]. The average values for violating the minimum inventory

Table 1. General results for routing and operational costs without penalization.

| Instance | BKS | Hybrid VNS | | | | |
|---|---|---|---|---|---|---|
| | | *min* | *avg* | *max* | *sd* | *time* |
| 20170809 | 2.83 | 1.47 | 1.60 | 1.87 | 0.11 | 1278 |
| 20170910 | 2.71 | 1.05 | 1.14 | 1.18 | 0.04 | 1383 |
| 20171011 | 2.57 | 0.98 | 1.02 | 1.09 | 0.03 | 1574 |
| 20171112 | 2.73 | 1.43 | 1.48 | 1.54 | 0.03 | 1176 |
| 20171201 | 2.42 | 1.08 | 1.24 | 1.44 | 0.12 | 1337 |
| 20180102 | 2.65 | 0.78 | 0.89 | 1.05 | 0.08 | 1506 |
| 20180203 | 2.06 | 0.94 | 1.05 | 1.14 | 0.07 | 1427 |
| 20180304 | 1.80 | 0.79 | 0.85 | 0.96 | 0.06 | 1413 |
| 20180405 | 1.93 | 0.65 | 0.71 | 0.80 | 0.05 | 1428 |
| 20180506 | 2.23 | 1.10 | 1.14 | 1.28 | 0.05 | 1408 |

limit are 119.33 for the BKS and 6.30 for the VNS. Regarding maximum limits, these values are 71.99 and 2.88, respectively.

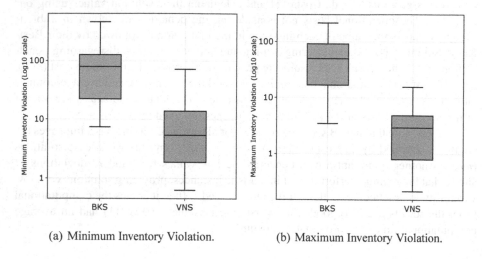

(a) Minimum Inventory Violation.          (b) Maximum Inventory Violation.

**Fig. 2.** Boxplots for the Violations distributions of each method.

Regarding the violation of draft constraints, the hybrid VNS also performed better than the solutions provided by da Costa [5]. For the maximum limit ($L_{iv}^{MX}$), the VNS violated it in only 1% of the runs with an average violation value of 0.33 against 30% of the runs and 7.27 on the average violation value for the BKS. Regarding the minimum limit ($L_{iv}^{MN}$), these values are 10% and 2.00 for the VNS, and 90% and 16.40 for the BKS.

## 6    Conclusions

The problem of routing ships with stock is widely studied in the literature, with different extensions of the basic case being considered [3], as well as different methods of solution. Through the literature review, it was noticed that there are opportunities for improving the solution methods and extending formulations to consider realistic aspects of the problem.

We achieved good results for a real-life multi-product maritime inventory routing problem (Multi-product MIRP), applied to the oil and gas industry, through a hybrid metaheuristic approach, combining a VNS framework with a linear mathematical model. The developed method applies random movements to modify the routing solution, running the mathematical formulation to optimize the inventory levels iteratively in the search for better solutions.

Experiments were conducted in a set of ten real-life instances with considerably large size, with nine products, 18 ports, and a fleet size ranging between 12 and 15 ships. A relevant aspect is a difficulty of finding feasible solutions to these instances.

Another interesting characteristic regards the fact that some ports do not have mandatory demands, while in other cases, products can be replaced by others with similar or higher quality without affecting the cost of the solution, giving more opportunity to meet the demands in some ports. Our approach presents good results with low variability, showing to be useful in practice. To evaluate our approach, we compared it with the solutions provided by da Costa [5]. All routes proposed by the author were reevaluated according to our criteria, making it possible to compare with the hybrid VNS. New best solutions were provided for all instances with fewer inventory violations and at a lower cost. Furthermore, computational times are within an acceptable range for the company's planning, allowing multiple re-planning on the same day.

There are still aspects not covered in this work and maybe the object of study of future works, such as testing the approach on larger instances and on the stochastic version of the MIRP. Few studies take into account the various uncertainties associated with this problem [1]. In maritime transport, uncertainties are frequent and often related to climatic conditions, ship reliability, port delays, and others. These uncertainties impact the ship's operation, whether in navigation or operation times, and might cause undesirable effects such as higher planning costs, inefficient fleet usage, or higher stock violations.

# References

1. Agra, A., Christiansen, M., Delgado, A., Hvattum, L.M.: A maritime inventory routing problem with stochastic sailing and port times. Comput. Oper. Res. **61**, 18–30 (2015)
2. Bertazzi, L., Coelho, L.C., De Maio, A., Laganà, D.: A matheuristic algorithm for the multi-depot inventory routing problem. Transp. Res. Part E: Logist. Transp. Rev. **122**, 524–544 (2019)
3. Christiansen, M., Fagerholt, K.: Maritime inventory routing problems. Encycl. Optim. **2**, 1947–1955 (2009)
4. Christiansen, M., Fagerholt, K., Flatberg, T., Haugen, Ø., Kloster, O., Lund, E.H.: Maritime inventory routing with multiple products: a case study from the cement industry. Eur. J. Oper. Res. **208**(1), 86–94 (2011)
5. da Costa, L.G.V.: Matheurísticas para a Roteirização de Navios com Estoques e Múltiplos Produtos. Master's thesis, PUC-Rio (2018). in Portuguese
6. Dauzère-Pérès, S., et al.: Omya Hustadmarmor optimizes its supply chain for delivering calcium carbonate slurry to European paper manufacturers. Interfaces **37**(1), 39–51 (2007)
7. Diz, G., Hamacher, S., Oliveira, F.: Maritime inventory routing: a practical assessment and a robust optimization approach. Ph.D. thesis, PUC-Rio (2017)
8. Hansen, P., Mladenović, N.: Variable neighborhood search. In: Handbook of Metaheuristics, pp. 145–184 (2003)
9. Hemmati, A., Hvattum, L.M., Christiansen, M., Laporte, G.: An iterative two-phase hybrid matheuristic for a multi-product short sea inventory-routing problem. Eur. J. Oper. Res. **252**(3), 775–788 (2016)
10. Kirkpatrick, S., Gelatt, C.D., Vecchi, M.P.: Optimization by simulated annealing. Science **220**(4598), 671–680 (1983)
11. Munguía, L.M., Ahmed, S., Bader, D.A., Nemhauser, G.L., Shao, Y., Papageorgiou, D.J.: Tailoring parallel alternating criteria search for domain specific MIPs: application to maritime inventory routing. Comput. Oper. Res. **111**, 21–34 (2019)

12. Papageorgiou, D.J., Cheon, M.-S., Harwood, S., Trespalacios, F., Nemhauser, G.L.: Recent progress using matheuristics for strategic maritime inventory routing. In: Konstantopoulos, C., Pantziou, G. (eds.) Modeling, Computing and Data Handling Methodologies for Maritime Transportation. ISRL, vol. 131, pp. 59–94. Springer, Cham (2018). https://doi.org/10.1007/978-3-319-61801-2_3

13. Papageorgiou, D.J., Nemhauser, G.L., Sokol, J., Cheon, M.S., Keha, A.B.: MIRPLib-A library of maritime inventory routing problem instances: survey, core model, and benchmark results. Eur. J. Oper. Res. **235**(2), 350–366 (2014)

14. Siswanto, N., Essam, D., Sarker, R.: Solving the ship inventory routing and scheduling problem with undedicated compartments. Comput. Ind. Eng. **61**(2), 289–299 (2011)

15. Song, J.H., Furman, K.C.: A maritime inventory routing problem: practical approach. Comput. Oper. Res. **40**(3), 657–665 (2013)

16. Subramanian, A., Drummond, L.M.A., Bentes, C., Ochi, L.S., Farias, R.: A parallel heuristic for the vehicle routing problem with simultaneous pickup and delivery. Comput. Oper. Res. **37**(11), 1899–1911 (2010)

17. Uggen, K.T., Fodstad, M., Nørstebø, V.S.: Using and extending fix-and-relax to solve maritime inventory routing problems. Top **21**(2), 355–377 (2013). https://doi.org/10.1007/s11750-011-0174-z

# Simplicial Vertex Heuristic in Solving the Railway Arrival and Departure Paths Assignment Problem

Damir Gainanov[1], Nenad Mladenović[2], and Varvara Rasskazova[3(✉)]

[1] Ural Federal University, Yekaterinburg, Russia
`damir.gainanov@gmail.com`
[2] Khalifa University, Abu Dhabi, United Arab Emirates
`nenad@turing.mi.sanu.ac.rs`
[3] Moscow Aviation Institute, Moscow, Russia
`varvara.rasskazova@mail.ru`
`http://www.mi.sanu.ac.rs/nenad/`

**Abstract.** This paper considers a fast solving the practical problem in railway planning and scheduling. i.e., the problem of assigning given arrival and departure railway paths to routs. This problem is to execute as fully as possible the train traffic across the railway station, using a fixed amount of the resources. It appears that the problem may be solved by using any efficient maximum *Independent set* algorithm, which is known to be $\mathcal{NP}$–hard. On the other hand, *Simplicial vertex test* is known heuristic that gives good quality solutions on sparse graphs. So, for solving the maximum independent set on sparse graphs, we propose an efficient heuristic based on the extended simplicial vertex test.

**Keywords:** Path assignment problem · Maximum independent set · Railway logistics

## 1  Introduction

Railway logistics is a huge area for applications of numerous combinatorial and graph based models. The most important stages of railway control are transportations planning, and also management of the technical and energy resources including their logistics and assignment. Among best known results in solving railway control problems it should be mentioned papers [1,2], where there are presented useful surveys on different applied combinatorial and graph based approaches.

Assignment problems often occur in railway locomotives logistics. So, in the paper [3] it is presented optimization model for this class of railway control problems. In the paper [4] authors developed an optimization locomotives assignment model with using stochastic programming methods. The same locomotives assignment problem was solved by schedule theory in the paper [5] and also by

© Springer Nature Switzerland AG 2021
N. Mladenovic et al. (Eds.): ICVNS 2021, LNCS 12559, pp. 123–137, 2021.
https://doi.org/10.1007/978-3-030-69625-2_10

graph based methods in the paper [6]. But railway arrival and departure paths assignment problem is not known very well. So, we introduce in the present paper some special terms and original statement of this optimization problem.

Railway arrival and departure paths assignment problem is very important stage of transportation management. Indeed, an effective execution of this stage implies increasing of bandwidth of the network. At the same time it could also increase a correct execution of transportations plan and efficient of railway management from economical point of view.

So, for solving railway arrival and departure paths assignment problem we propose in the present paper a graph based approach. This approach allows to reduce this problem to the well known $\mathcal{NP}$-hard maximum independent set problem.

To solving maximum independent set problem there are numerous fast heuristics, including branch-and-bound methods and simplicial vertex test. Among the most effective algorithms for solving maximum independent set problem it should be mentioned papers [7–9] and [10]. In the present paper we propose a modified heuristic algorithm which combines approaches from [9] and [10]. Following initial terminology this heuristic was named as extended simplicial vertex test with interchange step.

The paper presents a detailed description of mathematical model for investigation of the applied problem under consideration and also heuristic algorithm for solving corresponding optimization problem. The conclusion contains the results of computational experiments with data obtained from real railway control problem.

## 2    Railway Arrival and Departure Paths Assignment Problem (ADPA Problem)

Let $\mathbb{P} = \{1, m\}$ be the set of arrival and departure paths (ADP) available for trains assignment. And let $\mathbb{T} = \{1, n\}$ be the set of trains, such that for each train there defined parameters of the form

$$i\colon (arr(i), dep(i), \mathrm{p}(i)), i \in \{1, n\}. \tag{1}$$

Here $arr(i), dep(i) \in [T_0, T]$ are arrival and departure times for the $i$-th train, which both belong to the planning period $[T_0, T]$, and $\mathrm{p}(i) \subseteq \mathbb{P}$ is the set of arrival and departure paths available for the assignment of the $i$-th train. Note, that regarding to real application of the trains assignment problem, such a set $\mathrm{p}(i)$ depends on direction of the $i$-th train route and also on topological scheme of the railway station.

Let us denote a phantom arrival and departure path as $p_0$, and let us assume that $p_0 \in \mathrm{p}(i)$ for any $i \in \{1, n\}$.

**Definition 1.** *A feasible assignment of the arrival and departure paths to the set of trains given by parameters of the form (1) with fixed value of the parameter $\Delta t(\Delta t > 0)$, is the map of the form*

$$f: \mathbb{T} \longrightarrow \mathbb{P} \cup \{p_0\} \tag{2}$$

*such that for any* $i, j: f(i) = f(j) = k, i, j \in \{1, n\}, k \in \{1, m\}$ *the following condition holds*

$$\begin{cases} dep(i) + \Delta t \leq arr(j) \ \ or \ dep(j) + \Delta t \leq arr(i), \\ k \in \mathrm{p}(i) \cap \mathrm{p}(j). \end{cases} \tag{3}$$

The condition (3) from the Definition 1 means, that in case of the assignment of both $i$-th and $j$-th trains to the same $k$-th arrival and departure path, there should be the time delay between trains not less than the value of the parameter $\Delta t$. So, using the Definition 1 we can formulate an optimisation problem on the arrival and departure paths assignment (ADPA Problem).

*Problem 1.* For the given $\mathbb{T}, \mathbb{P}, p_0$ and $\Delta t$, one need to find a map $f_{sol}$ such that

$$|\{i: f_{sol}(i) = p_0\}| = \min_{f \in \mathfrak{F}} |\{i: f(i) = p_0\}|,$$

where $\mathfrak{F}$ is the set of all feasible assignments of the form (2) which satisfy condition (3).

In other words, ADPA Problem is to assign the maximal possible number of trains to the given number of paths with satisfying the constraint (3) for trains assigned to the same path. To solve Problem 1 we propose a graph based approach to reduce an applied problem under consideration to the maximum independent set problem.

## 3 The Maximum Independent Set in the ADPA Problem

Let us consider an undirected graph $G = (V, E)$ such that each vertex of this graph corresponds to a pair of the form "train-path", that is

$$V = \{v_i: train_i\text{-}path_i, train_i \in \{1, n\}, path_i \in \mathrm{p}(train_i)\}, \tag{4}$$

where $i \in \left\{1, \sum_{j \in \{1,n\}} |\mathrm{p}(j)|\right\}$. And let each pair of vertices $v_i, v_j$ are connected by edge $(v_i, v_j) \in E$ if the following condition holds

$$train_i = train_j, \tag{5}$$

or

$$\begin{cases} train_i \neq train_j, \\ path_i = path_j, \\ dep(i) + \Delta t > arr(j) \ \ or \ dep(j) + \Delta t > arr(i). \end{cases} \tag{6}$$

**Definition 2.** *An arrival and departure paths assignment graph (ADPA graph) is an undirected graph $G = (V, E)$, which satisfies conditions (4–6).*

We will explain rules (4)–(6) using simple instance.

*Example 1.* Let $\mathbb{P} = \{1, 2\}$ and $\mathbb{T} = \{1, 2, 3, 4\}$, such that

$$\mathbb{T} = \begin{cases} 1: (2:16, 2:43, \{1,2\}) \\ 2: (1:54, 2:20, \{1,2\}) \\ 3: (2:43, 3:14, \{1,2\}) \\ 4: (0:48, 1:16, \{1,2\}) \end{cases} \text{ and } \mathbb{T}' = \begin{cases} 1: (0:48, 1:16, \{1,2\}) \\ 2: (1:54, 2:20, \{1,2\}) \\ 3: (2:16, 2:43, \{1,2\}) \\ 4: (2:43, 3:14, \{1,2\}) \end{cases},$$

where $\mathbb{T}'$ is obtained by sorting of the set $\mathbb{T}$ regarding to values of parameter $arr(i)$. According to (4) from the set $\mathbb{T}'$ we have

$$V = \{\text{1-1}, \text{1-2}, \text{2-1}, \text{2-2}, \text{3-1}, \text{3-2}, \text{4-1}, \text{4-2}\}.$$

Using the value of the parameter $\Delta t = 0:10$ (that is $\Delta t$ will be equal to 10 min), from (5) and (6) we have the following set of edges

$$E = \{ (\text{1-1}, \text{1-2}), (\text{2-1}, \text{2-2}), (\text{3-1}, \text{3-2}), (\text{4-1}, \text{4-2}),$$
$$(\text{2-1}, \text{3-1}), (\text{2-2}, \text{3-2}), (\text{3-1}, \text{4-1}), (\text{3-2}, \text{4-2}) \}.$$

Thus, the adjacent matrix of the ADPA graph will have the form of Table 1.

Table 1. Adjacent matrix of the ADPA graph.

|     | 1-1 | 1-2 | 2-1 | 2-2 | 3-1 | 3-2 | 4-1 | 4-2 |
|-----|-----|-----|-----|-----|-----|-----|-----|-----|
| 1-1 | 1   | 1   | 0   | 0   | 0   | 0   | 0   | 0   |
| 1-2 | 1   | 1   | 0   | 0   | 0   | 0   | 0   | 0   |
| 2-1 | 0   | 0   | 1   | 1   | 1   | 0   | 0   | 0   |
| 2-2 | 0   | 0   | 1   | 1   | 0   | 1   | 0   | 0   |
| 3-1 | 0   | 0   | 1   | 0   | 1   | 1   | 1   | 0   |
| 3-2 | 0   | 0   | 0   | 1   | 1   | 1   | 0   | 1   |
| 4-1 | 0   | 0   | 0   | 0   | 1   | 0   | 1   | 1   |
| 4-2 | 0   | 0   | 0   | 0   | 0   | 1   | 1   | 1   |

The colored elements in the table above correspond to edges, which satisfy condition (5). From applied point of view it means that some fixed train could not be assigned to several different paths.

For further explanation we need the following additional result from graph theory. Let $\langle \tilde{V} \rangle_G$ be an induced subgraph of a graph $G = (V, E)$, which was generated by a subset of vertices $\tilde{V} \subseteq V$. That is

$$\langle \tilde{V} \rangle_G = \left( \tilde{V}, \tilde{E} \right) : \begin{cases} \tilde{V} \subseteq V, \\ \tilde{E} \subseteq E, \\ (u, v) \in \tilde{E} \text{ if and only if } u, v \in \tilde{V}, \end{cases}$$

where $u, v \in V, (u, v) \in E$.

**Proposition 1.** *Let $V$ be a set of vertices of a graph $G = (V, E)$, and let $V$ could be represented by $n$ subsets $V_i \subseteq V$, such that*

$$V = \bigcup_{i=1}^{n} V_i, V_i \subseteq V.$$

*Let also for any $i \in \{1, n\}$ the corresponding induced subgraph $\langle V_i \rangle_G$ is a complete subgraph (in other words, let for any $i \in \{1, n\}$ the corresponding subset of vertices $V_i$ is a clique of the initial graph). Then*

$$|\text{MaxIndSet}(G)| \leq n, \tag{7}$$

*where $\text{MaxIndSet}(G)$ is the maximum independent set of the graph $G = (V, E)$.*

*Proof.* As input we have that each subgraph of the form $\langle V_i \rangle_G$ is complete. So, for any $i \in \{1, n\}$ and for any pair of vertices $u, v \in V_i$ there holds either

$$\text{if } u \in \text{MaxIndSet}(G), \text{ then } v \notin \text{MaxIndSet}(G),$$

or

$$\text{if } v \in \text{MaxIndSet}(G), \text{ then } u \notin \text{MaxIndSet}(G).$$

Let us continue the proof using a mathematical induction method.

1. Let $n = 1$. Then $G = (V, E)$ is a complete undirected graph and

$$|\text{MaxIndSet}(G)| = 1 = n.$$

2. Let $n = k$ and $k > 1$. Following mathematical induction method we suppose that

$$|\text{MaxIndSet}(G)| \leq k.$$

3. Let $n = k + 1$. Let us denote

$$S = \text{MaxIndSet}\left(\langle \bigcup_{i=1}^{k} V_i \rangle_G\right)$$

and consider all possible cases.

Case 1

Let for any vertex $u \in V_{k+1}$ there is a vertex $v \in S$ such that either $(u, v) \in E$ or $(v, u) \in E$. Then according to induction step 2, we have

$$|\text{MaxIndSet}(G)| = |S| \leq k.$$

Case 2

Let there exists a vertex $u \in V_{k+1}$ such that for any vertex $v \in S$ we have both $(u, v) \notin E$ and $(v, u) \notin E$. Then

$$|\text{MaxIndSet}(G)| = |S| + 1 \leq k + 1 = n.$$

Thus, the proposition is proven for $n = 1$ and for $n = k + 1$ where $k > 1$. So, according to induction, the proposition holds for any $n$.

Note, that upper bound of the form (7) holds for any set of cliques of the graph. In particular, subsets $\langle V_i \rangle_G, i \in \{1, n\}$ could be intersecting to each other by vertices (or edges).

Now we will return to explanation of the graph based approach to solve ADPA Problem 1. We state that any independent set of vertices of the ADPA graph $G = (V, E)$ corresponds to arrival and departure paths assignment as following rule

$$v_i : train_i\text{-}path_i \iff f(train_i) = path_i, \tag{8}$$

where $v_i \in V, train_i \in \{1, n\}, path_i \in \{1, m\}$.

**Proposition 2.** *The maximum independent set of the ADPA graph $G = (V, E)$ corresponds to optimal solution of the ADPA Problem 1.*

*Proof.* Using (5) one can conclude that any ADPA graph $G = (V, E)$ contains $n$ complete subgraphs of the form $G_k = (V_k, E_k)$, such that

$$V_k = \{v_i \in V : train_i = k\}, |E_k| = |p(k)|,$$

and

$$\bigcup_{k \in \{1, n\}} V_k = V,$$

where $V_k \subseteq V, E_k \subseteq E$ and $p(k) \subseteq \{1, m\}$ for any $k \in \{1, n\}$. In other words, condition of Proposition 1 holds for any ADPA graph $G = (V, E)$.

Then, for any independent set $S$ of the ADPA graph $G = (V, E)$ the following holds

$$|S| \leq n, \tag{9}$$

and moreover, for any $v_i, v_j \in S$ either

$$\begin{cases} train_i \neq train_j, \\ path_i \neq path_j \end{cases}$$

or

$$\begin{cases} train_i \neq train_j, \\ path_i = path_j, \\ dep(i) + \Delta t \leq arr(j) \text{ or } dep(j) + \Delta t \leq arr(i). \end{cases}$$

Thus, it is right that for any independent set $S$ the corresponding rule (8) satisfies to condition (3) for trains assigned to the same arrival and departure path.

Further, if $|S| = n$, than the proposition is proved, and for corresponding assignment of the form (2) the following equality holds

$$|\{i : f(i) = p_0\}| = 0.$$

Let us suppose that $|S| \neq n$, and let there exists the assignment of the form (2), such that

$$|S| < |\{i\colon f(i) \neq p_0\}| \leq n.$$

In this case due to (6) and (8), corresponding vertices of the ADPA graph need to be not adjacent. However, this conclusion contradicts to maximality of the set $S$ by cardinality.

It should be also mentioned, that inequality (9) provides an additional stopping condition from algorithmic point of view.

## 4  Extended Simplicial Vertex Test with Interchange Step (ESVTwICS)

In the paper [11] there was developed an inference algorithm for monotone Boolean functions, generated by undirected graphs. In the paper [9] the algorithm mentioned above was applied to solving the problem on the maximum independent set of the undirected graph (MIS problem). Following terminology from [12] this algorithm was named as extended simplicial vertex test (ESVT) (thanks to some vertices features used as selecting criterion for candidates to be included in the solution). In the paper [13] the ESVT was effectively used for solving an applied problem on the railway management on the stage of forming the set of feasible trains schedules.

In the paper [10] there was developed an algorithm for solving the maximum clique problem, which is known to be equivalent to MIS Problem. One of steps of this algorithm is an interchange step. The meaning of interchange step is to exclude the random vertex from local solution and try to extend the last one using remained vertices, which were earlier not used for local solution construction. So, an interchange is a step of local search in frame of variable neighborhood metaheuristic (VNS).

Finally, the extended simplicial vertex test with interchange step (ESVTwICS) which we propose in the present paper for solving an applied problem, is a constructive heuristic algorithm which uses an interchange step for local solution obtained by ESVT.

For further explanation we need to introduce some additional notes and definitions. Let

$$\mathcal{N}(v, V, E) = \{u\colon u, v \in V, (u, v) \in E\}$$

be a neighborhood of vertex $v \in V$ in a graph $G = (V, E)$, that is a subset of vertices from $V$ which are adjacent to vertex $v$ by edges from $E$. And let

$$\mathcal{A}(v, V, E) = \{(u, w) \in E\colon u, w \in \mathcal{N}(v, V, E)\}$$

be a subset of edges $(u, w) \in E$ of a graph $G = (V, E)$ which both ends belong to a neighborhood of vertex $v \in V$.

**Definition 3.** *Parameters $k$ and $m$ of a vertex $v \in V$ of a graph $G = (V, E)$ are the following values*

$$k = |\mathcal{N}(v, V, E)|,$$

$$m = \frac{k \cdot (k-1)}{2} - |\mathcal{A}(v, V, E)|$$

*corresponding.*

In other words, a parameter $k$ is a number of neighbors of a vertex $v$ and a parameter $m$ is a number of edges missed in the neighborhood of this vertex to be a complete induced subgraph. Note, that a vertex with parameter $m = 0$ is a simplicial vertex by definition from [10].

A figure below shows a general scheme of ESVTwICS (Fig. 1).

---

**Require:** $G = (V, E), n$       ▷ an undirected graph and an integer number $n$
**Ensure:** $S \subset V$       ▷ an independent set of vertices
1:   $V_0 \longleftarrow V$       ▷ current set of candidates
2:   $E_0 \longleftarrow E$
3:   $S \longleftarrow \{\}$       ▷ an independent set under construction
4:   **while** $V_0 \neq \{\}$ **do**       ▷ ESVT step
5:      **for all** $v \in V_0$ **do**
6:         **for all** $u \in V_0$ **do**
7:            visited $(u) \longleftarrow false$
8:         $k(v) \longleftarrow$ CALCULATE-K$(v, V_0, E_0)$
9:         $m(v) \longleftarrow$ CALCULATE-M$(v, V_0, E_0)$
10:     $v_0 \longleftarrow$ MIN-MAX-PARAM$(k, m, V_0)$
11:     $S \longleftarrow S \cup \{v_0\}$
12:     $V_0 \longleftarrow V_0 \setminus \{v_0\} \setminus \mathcal{N}(v_0, V_0, E_0)$
13:     $E_0 \longleftarrow E_0 \setminus \{(v_0, u) : u \in V_0\} \setminus \{(u, w), (w, u) : u \in \mathcal{N}(v_0, V_0, E_0), w \in V_0\}$
14: **if** $|S| = n$ **then**       ▷ stopping condition
15:     **go to** step 21
16: **for all** $v \in S$ **do**       ▷ interchange step
17:     $S' \longleftarrow$ INTERCHANGE$(v, S, V, E)$
18:     **if** $|S'| > |S|$ **then**
19:        $S \longleftarrow S'$
20:        **go to** step 14
21: **return** $S$

---

**Fig. 1.** Algorithm ESVTwICS$(G, n)$.

As one can see from Fig. 1 the ESVT step (lines 4–13) calculates values of parameters $k$ and $m$ for vertices from current set $V_0$ on each iteration of an external loop. Pseudo-codes of corresponding procedures are shown by Fig. 2 and 3.

A boolean array visited() used in lines 6–7, Fig. 1 and lines 2, 5–6 and 8, Fig. 3 has a length $|V|$ and allows one to check only relevant vertices. Further, both

```
1: procedure CALCULATE-K(v, V, E)
2:    k (v) ⟵ 0
3:    for all u ∈ V do
4:        if (u, v) ∈ E then
5:            k (v) ⟵ k (v) + 1
6:    return k (v)
```

**Fig. 2.** Procedure CALCULATE-K$(v, V, E)$.

```
1: procedure CALCULATE-M(v, V, E)
2:    visited (v) ⟵ true
3:    m (v) ⟵ 0
4:    for all u ∈ V do
5:        if visited (u) = false & (u, v) ∈ E then
6:            visited (u) ⟵ true
7:            for all w ∈ V do
8:                if visited (w) = false & (v, w) ∈ E & (u, w) ∉ E then
9:                    m (v) ⟵ m (v) + 1
10:    return m (v)
```

**Fig. 3.** Procedure CALCULATE-M$(v, V, E)$.

```
1: procedure MIN-MAX-PAR(k, m, V)
2:    min-m ⟵ 0
3:    max-k ⟵ 0
4:    ind ⟵ false
5:    while ind ⟵ false do
6:        for all v ∈ V do
7:            if m (v) = min-m & k (v) > max-k then
8:                max-k ⟵ k (v)
9:                v₀ ⟵ v
10:               ind ⟵ true
11:       if ind=false then
12:           min-m ⟵ min-m + 1
```

**Fig. 4.** Procedure MIN-MAX-PAR$(k, m, V)$.

integer arrays $k()$ and $m()$ also have a lengths $|V|$ and store a data regarding to values of corresponding values of parameters $k$ and $m$ for each vertex from the current set.

The main idea of ESVT is to add to a solution under construction $(S)$ a vertex $v$ with minimum value of the parameter $m$ and the maximum one of the parameter $k$. So, let's describe corresponding procedure to extracting a vertex with features mentioned above.

Procedure MIN-MAX-PAR($k, m, V$) uses a boolean variable ind to check a stopping condition of an external loop (line 5, Fig. 4). By the other hand, if this stopping condition is not executed one need to increase a value of parameter $m$ (line 10, Fig. 4). Note, that an original simplicial vertex test (not extended) use only criterion $m = 0$.

Finally, the last procedure which we need to discuss is an interchange (lines 16–20, Fig. 1). The figure below shows a pseudocode of this procedure.

---

```
1: procedure INTERCHANGE(v, S, V, E)
2:     S' ⟵ S \ {v}
3:     V' ⟵ V \ S
4:     for all u ∈ S' do
5:         for all w ∈ V' do
6:             if (u, w) ∈ E then
7:                 V' ⟵ V' \ {w}
8:     if |V'| > 1 then
9:         for all u ∈ V' do
10:            visited (u) ⟵ true
11:            for all w ∈ V' do
12:                if visited (w) = false & (u, w) ∉ E then
13:                    S' ⟵ S' ∪ {u, w}
14:                    break
15:     return S'
```

---

**Fig. 5.** Procedure INTERCHANGE($v, S, V, E$).

As earlier (Fig. 2 and 3) an interchange procedure also uses boolean array visited() to check only relevant vertices. So, we declare all vertices from $V'$ as unvisited (lines 6–7, Fig. 5) to run interchange correctly (lines 16–20, Fig. 1) in case of multiple checking of stopping condition (lines 18–20, Fig. 1).

The main idea of interchange is to exclude 1 vertex from local solution $S$ (line 2, Fig. 5) and change it with 2 another vertices. For this goal one need to construct a special set $V'$ (line 3, Fig. 5) which contains the only vertices adjacent to this excluded vertex. Note, that a candidate $v$ for excluding could be chosen at random or using some heuristic rule. For example, a candidate could have a minimal degree in an initial graph.

Next step of interchange (lines 8–9, Fig. 5) is to check, whether there are in constructed set $V'$ a pair of vertices, which are not adjacent with each other. If so, then each of them is not adjacent with other vertices from $S'$ and could be added to independent set under construction (line 13, Fig. 5). Thus, one can change 1 (excluded) vertex from local solution with 2 another vertices. The same principals could be used in case of change 2 vertices with 3 ones and so on.

For more efficiency we use interchange for all vertices from local solution obtained by ESVT (line 16, Fig. 1).

*Example 2.* Let us consider an undirected graph $G = (V, E)$ shown by Fig. 6.

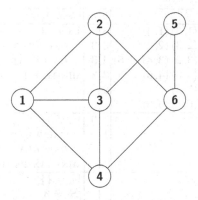

**Fig. 6.** An undirected graph $G = (V, E)$.

Using Proposition 1 one can state $n = 3$ as a stopping condition for the ESVTwICS. Indeed, a set of vertices of a given graph could be represented by $V = \{1, 2, 3\} \cup \{1, 3, 4\} \cup \{5, 6\}$, where each of subsets $\{1, 2, 3\}$, $\{1, 3, 4\}$, $\{5, 6\}$ is a clique of the initial graph. So, we have $|\text{MaxIndSet}\,(G)| \leq 3$ and $n = 3$.

Implementation of ESVTwICS with instance $G = (V, E)$ (Fig. 6) is represented by Table 2 below.

First of all we run ESVT for given set of vertices and then calculate parameters $k$ and $m$ for each vertex in view of the given set of edges. The loop run until the current set of vertices becomes empty. So, we have got an independent set, which contains 2 vertices. But this independent set does not satisfies the stopping condition, that's why the algorithm go to interchange step. Running interchange twice for each vertex from local solution obtained by the ESVT step, we are getting an optimal one, for which the stopping condition holds.

## 4.1  Computational Results

We run ESVTwICS with instances obtained from real trains schedule in the period from Jan, 1st until Jan, 30th. Note, that as usually each train has its own calendar plan. That is why for different dates we have different number of transportations across the station. At the same time less number of transportations does not imply any simplification of the model. Indeed, although number of transportations is not very large, but it could be often occur that trains schedule are very close to each other and in this case it is impossible to execute arrival and departure paths assignment without changing the trains schedule.

Let us describe assumptions of the model.

1. $\Delta t = 10$ (min). It is a usual management condition on the delay between trains assigned to the same arrival and departure path.

**Table 2.** Running of the ESVTwICS $(G, n)$.

| ESVT step | | | | | |
|---|---|---|---|---|---|
| $S$ | $V_0$ | $E_0$ | $k$ | $m$ | Comments |
| {} | 1 | $(1,2)\,(1,3)\,(1,4)$ | 3 | 1 | $V_0 \neq \{\}$ |
|  | 2 | $(2,1)\,(2,3)\,(2,6)$ | 3 | 3 | CALCULATE-K $(v, V_0, E_0)$ |
|  | 3 | $(3,1)\,(3,2)\,(3,4)\,(3,5)$ | 4 | 3 | CALCULATE-M $(v, V_0, E_0)$ |
|  | 4 | $(4,1)\,(4,3)\,(4,6)$ | 3 | 3 | MIN-MAX-PAR $(k, m, V_0)$ |
|  | 5 | $(5,3)\,(5,6)$ | 2 | 1 |  |
|  | 6 | $(6,2)\,(6,4)\,(6,5)$ | 3 | 2 |  |
| {1} | 5 | $(5,6)$ | 1 | 0 | $V_0 \neq \{\}$ |
|  | 6 | $(6,5)$ | 1 | 0 | CALCULATE-K $(v, V_0, E_0)$ |
|  |  |  |  |  | CALCULATE-M $(v, V_0, E_0)$ |
|  |  |  |  |  | MIN-MAX-PAR $(k, m, V_0)$ |
| $\{1,5\}$ | {} |  |  |  | $V_0 = \{\}$ |
|  |  |  |  |  | $|S| \neq 3$ |

| Interchange step | | | | | | |
|---|---|---|---|---|---|---|
| $S$ | $S'$ | $V'$ | $v$ | $u$ | $w$ | Comments |
|  |  |  | 5 |  |  | **for all** $u \in S'$ **do** |
| $\{1,5\}$ | $\{1\}$ | $\{2,3,4,6\}$ |  | 1 |  | **for all** $w \in V'$ **do** |
|  |  |  |  |  | 2 | $(1,2) \in E$ |
|  |  | $\{3,4,6\}$ |  |  | 3 | $(1,3) \in E$ |
|  |  | $\{4,6\}$ |  |  | 4 | $(1,4) \in E$ |
|  |  | $\{6\}$ |  |  | 6 | $(1,6) \notin E$ |
|  |  | $\{6\}$ |  |  |  | $|V'| = 1$ |
|  |  |  |  |  |  | **return** $S'$ |
| $\{1,5\}$ | $\{5\}$ |  | 1 |  |  |  |
|  |  | $\{2,3,4,6\}$ |  | 5 | 2 | $(5,2) \notin E$ |
|  |  | $\{2,3,4,6\}$ |  |  | 3 | $(5,3) \in E$ |
|  |  | $\{2,4,6\}$ |  |  | 4 | $(5,4) \notin E$ |
|  |  | $\{2,4,6\}$ |  |  | 6 | $(5,6) \in E$ |
|  |  | $\{2,4\}$ |  |  |  | $|V'| > 1$ |
|  |  |  |  | 2 |  | **for all** $u \in V'$ **do** |
|  |  |  |  |  | 2 | visited $(2) = true$ |
|  |  |  |  |  |  | **for all** $w \in V'$ **do** |
|  |  |  |  |  | 4 | visited $(4) = false$ & $(2,4) \notin E$ |
|  |  | $\{2,4,5\}$ |  |  |  | **return** $S'$ |
| $\{1,5\}$ |  |  |  |  |  | $|S'| > |S|$ |
| $\{2,4,5\}$ |  |  |  |  |  | $S \longleftarrow S'$ |
|  |  |  |  |  |  | $|S| = 3$ |

2. $\Delta t = 20$ (min). We use such value of the parameter $\Delta t$ to demonstrate a stability of the approach proposed. And moreover it could be rather useful in view of possible deviations in real execution of transportations plan.

3. $\mathbb{P} = \{1, 2, 3, 4, 5, 6\}$—the set of arrival and departure paths available for assignment. We use only 6 paths instead of 9 existing, to get stronger constraints for ADPA Problem.

4. $p(i) = \mathbb{P}$ for any train during any date. This is the basic case which can often occur in real stations, when some paths maintenance actions need to be organized.

The Table 3 below shows the computation results which we get by using ESVTwICS with assumptions described.

**Table 3.** Computational results

| Instance | $\Delta t = 10$ | $\Delta t = 20$ | Instance | $\Delta t = 10$ | $\Delta t = 20$ | Instance | $\Delta t = 10$ | $\Delta t = 20$ |
|---|---|---|---|---|---|---|---|---|
| 70 | 70 | 69 | 72 | 72 | 69 | 70 | 70 | 67 |
| 78 | 77 | 74 | 80 | 80 | 76 | 76 | 76 | 72 |
| 72 | 72 | 70 | 72 | 72 | 68 | 68 | 68 | 67 |
| 77 | 76 | 74 | 76 | 75 | 73 | 72 | 72 | 70 |
| 74 | 74 | 71 | 68 | 68 | 67 | 72 | 72 | 69 |
| 79 | 78 | 75 | 78 | 77 | 73 | 80 | 80 | 76 |
| 70 | 70 | 67 | 70 | 70 | 68 | 72 | 72 | 68 |
| 76 | 76 | 72 | 77 | 76 | 74 | 76 | 75 | 73 |
| 68 | 68 | 67 | 74 | 74 | 71 | 68 | 68 | 67 |
| 72 | 72 | 70 | 79 | 78 | 75 | 78 | 77 | 74 |

In columns "Instance" there are shown numbers of trains, which need to be assigned to 6 paths (e.g., for each line of the Table 3 the corresponding ADPA graph will contain "number of trains" ×6 vertices). In both columns "$\Delta t = 10$" and "$\Delta t = 20$" there are presented the cardinalities of independent sets obtained by ESVTwICS run for corresponding ADPA graphs with given values of the parameter $\Delta t$.

As it could be seen from Table 3, in many cases ESVTwICS allows to assign all trains to given paths. In particular, for 21 instances among 30 there were obtained an exact solutions with using 10 min as a value of the parameter $\Delta t$.

Thus, observing results obtained by ESVTwICS we can conclude its efficiency to solving real problems regarding to station paths resources assignment and maintenance planning.

## 5   Conclusion

The paper proposes an efficient graph based approach to solving a railway arrival and departure paths assignment problem. There was developed a mathematical model which allow to reduce an applied problem to the classical $\mathcal{NP}$-hard maximum independent set problem. There are also described rules for constructing a special graph connected with applied input data. This graph was named as ADPA graph and serve as instance for modified simplicial vertex test heuristic (ESVTwICS).

The ESVTwICS heuristic use two extensive steps—extended simplicial vertex test and interchange step.

The paper also presents computational results with using real railway transportations schedule. These results show the efficiency of the approach proposed in railway management and logistics.

The future researches in the discussed direction will be devoted to combinatorial comparison of different graph based models in frame of their application to solving railway arrival and departure paths assignment problem. At the same time it seems to be interesting to increase the complexity of the problem by using additional constraints such as different types of trains and paths available for assignment.

As for direction of algorithms development for solving ADPA, it seems to be efficient to implement VNS approach using 2 neighborhoods described in the paper. Such a way has the highest priority due to its powerful in solving different problems regarding to transportations and assignments.

# References

1. Lazarev, A., Pravdivets, N., Nekrasov, I.: Evaluating typical algorithms of combinatorial optimization to solve continuous-time based scheduling problem. Algorithms **11**(4), 1–13 (2018)
2. Matyukhin, V., Shabunin, A., Kuznetsov, N., Takmazian, A.: Rail transport control by combinatorial optimization approach. In: 11th IEEE International Conference on Application of Information and Communication Technologies, pp. 419–422. IEEE Communications Society (2017)
3. Piu, F., Speranza, M.: The locomotive assignment problem: a survey on optimization models. Int. Trans. Oper. Res. **21**, 327–352 (2014)
4. Ivanov, S.V., Kibzun, A.I., Osokin, A.V.: Stochastic optimization model of locomotive assignment to freight trains. Autom. Remote Control **77**(11), 1944–1956 (2016). https://doi.org/10.1134/S0005117916110059
5. Lazarev, A.A., Musatova, E.G.: The problem of trains formation and scheduling: integer statements. Autom. Remote Control **74**(12), 2064–2068 (2013)
6. Azanov, V.M., Buyanov, M.V., Gaynanov, D.N., Ivanov, S.V.: Algorithm and software development to allocate locomotives for transportation of freight trains. Bull. South Ural State Univ. Ser. Math. Model. Program. Comput. Softw. **9**(4), 73–85 (2016)
7. Tomita, E., Imamatsu, K., Kohata, Y., Wakatsuki, M.: A simple and efficient branch and bound algorithm for finding a maximum clique with experimental evaluations. Syst. Comput. Japan **28**, 60–67 (1997)
8. Abello, J., Pardalos, P.M., Resende, M.: On maximum clique problems in very large graphs. In: External Memory Algorithms, pp. 119–130 (1999)
9. Gainanov, D., Mladenović, N., Rasskazova, V., Urošević, D.: Heuristic algorithm for finding the maximum independent set with absolute estimate of the accuracy. In: CEUR-Workshop Proceedings, vol. 2098, pp. 141–149 (2018)
10. Hansen, P., Mladenović, N., Urošević, D.: Variable neighborhood search for the maximum clique. Discret. Appl. Math. **145**(1), 117–125 (2004)
11. Gainanov, D., Rasskazova, V.: An inference algorithm for monotone Boolean functions associated with undirected graphs. Bull. South Ural State Univ. Ser. Math. Model. Program. Comput. Softw. **9**(3), 17–30 (2016)

12. Hertz, A., Friden, C., De Werra, D.: Tabaris: an exact algorithm based on tabu search for finding a maximum independent set in a graph. Comput. Oper. Res. **17**, 437–445 (1990)
13. Gainanov, D., Mladenović, N., Rasskazova, V.: Maximum independent set in planning freight railway transportation. Front. Eng. Manag. **5**(4), 499–506 (2018)

# Scheduling of Patients in Emergency Departments with a Variable Neighborhood Search

Thiago Alves de Queiroz[1]([✉]) [iD], Manuel Iori[2] [iD], Arthur Kramer[3] [iD], and Yong-Hong Kuo[4] [iD]

[1] Institute of Mathematics and Technology, Federal University of Catalão, Catalão, GO 75704-020, Brazil
taq@ufg.br
[2] Department of Sciences and Methods for Engineering, University of Modena and Reggio Emilia, 42122 Reggio Emilia, Italy
manuel.iori@unimore.it
[3] Department of Production Engineering, Federal University of Rio Grande do Norte, Natal, RN 59077-080, Brazil
arthur.kramer@ct.ufrn.br
[4] Department of Industrial and Manufacturing Systems Engineering, The University of Hong Kong, Pokfulam Road, Pok Fu Lam, Hong Kong
yhkuo@hku.hk

**Abstract.** The dynamic scheduling of patients to doctors in an emergency department environment is tackled in this work. We consider the case in which patients arrive dynamically during the working hours, and the objective is to minimize the weighted tardiness. We propose a greedy heuristic based on priority queues and a general variable neighborhood search (GVNS). In the greedy heuristic, patients are scheduled by observing their urgency, while in the GVNS, the schedule is optimized every time a patient arrives. The GVNS uses six neighborhood structures and a variable neighborhood descent to perform the local search. The GVNS also handles the static problem whose solution can be used as a reference for the dynamic one. Computational results on 80 instances show that using the GVNS better approximates the static problem, besides giving an overall reduction of 66.8% points over the greedy heuristic.

**Keywords:** Dynamic scheduling · Health care · Emergency department · Variable neighborhood search

## 1 Introduction

Emergency Department (ED) overcrowding has been continuously reported for decades in various regions in the world [6,26,28]. Since ED is a 24/7 gateway to the hospital for patients who require immediate emergency medical services, congestion within the facility may prevent those patients from accessing the

N. Mladenovic et al. (Eds.): ICVNS 2021, LNCS 12559, pp. 138–151, 2021.
https://doi.org/10.1007/978-3-030-69625-2_11

required treatments promptly. Such delays in the necessary medical treatments can lead to life-altering (or even life-ending) cases.

ED overcrowding's adverse consequences include public safety at risk, prolonged suffering, patient dissatisfaction, violence at the ED waiting room, and increased chances of decision errors [5]. One of the most effective ways to alleviate the ED overcrowding situation is to expand the facility's capacity and resources. This may not be feasible for most hospitals, particularly for public hospitals, due to financial constraints. Therefore, hospital management and operations managers have kept investigating possible ways to improve patient flow by optimizing the processes and services of the ED.

The flow of patients and the prolonged patient waiting times within EDs have always been studied and discussed. This macro-topic is relevant to various optimization problems, ranging from the scheduling of work shifts to the minimization of service costs. Our research is motivated by the recent advance in information technologies adopted by EDs. By modern information systems at the hospital, most of the ED activities are now tracked in real-time for more effective communications and responsive actions. Traditionally, as information might not be updated and comprehensive, protocols were set up to guide daily operations (e.g., patient prioritization, professional medical assignment, and process flow).

A conventional and commonly adopted practice is prioritizing patients according to the level of urgency or patient criticality. While this is a sensible way of giving priority to patients, the utilization of additional information about patients and resource availability may potentially further enhance the efficiency of the ED and effectiveness of emergency medical services provided to patients, thereby leading to better patient outcomes. For example, studies are suggesting that taking patient complexity into account would benefit ED performance [27].

The intervention strategies suggested by research work and adopted in practice for patient prioritization is mainly based on the triage category [8,29]. The more urgent a patient, the higher the priority is. [19] adopted a simulation approach to examine the impacts of the adoption of a fast track. They found that the fast track is more effective for EDs with a higher proportion of urgent patients. They provided conditions for which this scheme is incredibly useful. However, recent research found that patient complexity is still not a significant factor considered in patients' prioritization decisions [7].

Scheduling has been an essential topic in the optimization of ED systems. The leading research is on staff scheduling. Because of the stochastic environment of EDs, queuing and simulation models have been incorporated into optimization frameworks. For example, [12] used a Lag stationary independent period-by-period queuing model to determine the number of service providers in each hour. On the other hand, simulation-optimization approaches require the execution of simulation models at each iteration of an optimization procedure and, therefore, are computationally expensive [18,20].

Our work addresses the ED patient scheduling problem, where information about the patient characteristics and doctors' availability are provided. With such useful information, the scheduling of patients in the ED can be more productive and dynamic. More specifically, this paper focuses on minimizing the patients' waiting time by optimizing patients' schedules to doctors. The problem is further complicated by the heterogeneous patients presenting to the ED, characterized by the urgency and time required for medical treatments. The problem is dynamic, where no information about the patients is known before they arrive in the ED. Then, we propose a greedy heuristic based on priority queues to schedule patients. A General Variable Neighborhood Search (GVNS) [13] is used to optimize the schedule whenever a new patient arrives. We also solve the static version of the problem with the GVNS once it has all information available at time zero, and an ideal solution would be expected.

Despite the relevance of scheduling problems, which have been studied since the 1950s, the number of works addressing the same problem here is minimal. To our knowledge, the paper by [14] is the most recent work dedicated to our problem. These authors investigated dominance rules and proposed a branch-and-bound algorithm. In [1], the problem and its unweighted versions were studied. In addition to these works, we highlight the contributions by [15], who proposed a metaheuristic and provided an annotated bibliography review, and [2], who developed a branch-cut-and-price algorithm.

Concerning the use of the VNS to solve scheduling problems, a resource-constrained project scheduling was solved in [9]. A solution was coded as a sequence of valid activities in terms of precedence constraints. The nurse rostering problem was solved with a basic VNS combined with an integer programming model in [3], considering real instances arising in a Dutch hospital. A hybrid flow shop problem was tackled in [22] with a hybrid VNS that combines the chemical-reaction optimization and the estimation of distribution methods. In [4], the problem of scheduling surgeries over a medium-term horizon was solved with a VNS, outperforming a commercial solver for many real instances of a Dutch cardiothoracic center. On the other hand, the home healthcare routing and scheduling problems were tackled in [10] with a general VNS whose solutions could outperform other literature approaches. Recently, a variable neighborhood descent was used to handle the bicriteria parallel machine scheduling problem in [25], where a set of neighborhood structures based on swap, remove, and insertion moves were used.

The contributions of our work are as follows.

1. We address a dynamic ED patient scheduling problem, which is not well-studied in the existing literature.
2. A priority queue-based greedy heuristic and a GVNS algorithm are proposed to provide fast and effective solutions, which are novel to the dynamic ED patient scheduling problem.
3. A new design of computational experiments that simulate an ED dynamic environment was adopted to demonstrate the GVNS approach's effectiveness.

The present work is organized as follows. Section 2 provides the problem definition, while Sect. 3 has the solutions methods we proposed. Section 4 shows the computational experiments. Section 5 brings the concluding remarks and draws perspectives for future works.

## 2   Problem Definition

We address a particular scheduling problem that arises in the ED context. In this problem, we are given a set $J = \{1, \ldots, n\}$ of patients to be served, without preemption, by a set $M = \{1, \ldots, m\}$ of parallel doctors with identical efficiency. Each patient $j \in J$ has a service time $p_j$, an arrival date at the ED $r_j$, a priority weight $w_j$, and a due date $d_j$. One doctor must serve each patient.

The objective is to find a schedule of patients such that the sum of weighted tardiness of the patients, i.e., $\sum_{j \in J} w_j T_j$, is minimized. The tardiness of a patient $j$ is defined as $T_j = \max\{0, C_j - d_j\}$, where the completion time $C_j$ is defined as the time at which the patient's service is finished. Using the scheduling classification of [11], this problem can be referred to as $P|r_j| \sum w_j T_j$. The $P|r_j| \sum w_j T_j$ is $\mathcal{NP}$-hard, because it is a generalization of the $1|r_j| \sum w_j C_j$ that was proven $\mathcal{NP}$-hard by [21].

When all data is deterministic and known in advance, we have the static version of the problem. On the other hand, in the dynamic version, data are made available only immediately after the arrival of a patient in the ED. Figure 1 shows an example of a problem instance of five patient arrivals. In the static version of the problem, scheduling decisions are made at time $t = 0$, with $\{(r_j, p_j, w_j, d_j)\}_{j=1}^{5}$ known in advance. In the dynamic version, decisions are determined sequentially at time points $t = r_1, r_2, \ldots, r_5$, where only the information about the patient arrivals prior to the current arrival is known. For example, at $t = r_3$, $\{(r_j, p_j, w_j, d_j)\}_{j=1}^{3}$ are known but $\{(r_j, p_j, w_j, d_j)\}_{j=4}^{5}$ are unknown.

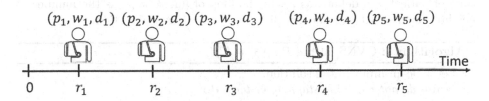

**Fig. 1.** An example of a problem instance of five patient arrivals.

In the next section, we describe a general variable neighborhood search for both the static and the dynamic versions, besides a greedy heuristic based on queues for the dynamic problem.

## 3   Solution Methods

The *variable neighborhood search* (VNS) was proposed by [23] and considers the systematical change of neighborhood structures to obtain a globally optimal solution for all neighborhood structures. It may include a local search phase.

When the local search is the *variable neighborhood descent* (VND), we have the *general* version of the VNS (GVNS). In the next subsections, we start describing the GVNS we proposed for the static problem. Next, the neighborhood structures we used in GVNS are explained. Finally, we discuss the proposed heuristics for the dynamic problem, including an adaptation of the GVNS.

## 3.1   General Variable Neighborhood Search

We describe the GVNS for the static version of the $P|r_j| \sum w_j T_j$ in Algorithm 1. As mentioned, this version assumes that information is known in advance. The GVNS has the following main steps: to create an initial solution randomly; to obtain neighbor solutions of the current solution by using a neighborhood structure; to do the local search with the VND on a neighbor solution; and, to accept the solution of the VND if it is better than the current solution. Whenever a solution is accepted, the GVNS restarts to the first neighborhood structure; otherwise, it goes to the next neighborhood.

In Algorithm 1, a solution $x$ is coded as a vector of lists of integers. Each vector position represents a machine (i.e., a doctor), so the vector has size $m$, and it contains an ordered list of jobs (i.e., patients). The initial solution is created as follows: for each patient $j \in J$, select randomly a doctor $m_t \in M$ and then assign $j$ to be the last patient of $m_t$. The start time of each patient $j$ is defined as the maximum between its release date $r_j$ and the completion time $C_i$ of its predecessor patient $i$ served by the same doctor. The value of a solution $x$, given by $f(x)$, is defined as $\sum_{j \in J} w_j T_j$, where $T_j$ is the patient's tardiness. The start time of patient $j$ is defined as $s_j$. In the loop of line 4, $k_{max}$ is the number of neighborhood structures that are discussed in the next subsection.

---

**Algorithm 1:** GVNS for the $P|r_j| \sum w_j T_j$.

---

1   Let $x$ be an input initial solution.
2   **while** *do not reach the stopping criteria* **do**
3   |   $k \leftarrow 1$.
4   |   **while** $k \leq k_{max}$ **do**
5   |   |   $x' \leftarrow$ obtain a random neighbor in $N_k(x)$.
6   |   |   $x'' \leftarrow \text{VND}(x')$.
7   |   |   **if** $f(x'') > f(x)$ **then**
8   |   |   |   $s \leftarrow s''$.
9   |   |   |   $k \leftarrow 1$.
10  |   |   **else**
11  |   |   |   $k \leftarrow k + 1$.

12  **return** $x$.

---

Regarding line 5 of Algorithm 1, the doctors and patients in the neighborhood structures are selected at random. On the other hand, in line 6, where the VND is called, we use the same neighborhood structures, but now instead of random selections, all the possibilities for doctors and patients are explored. The local search is described in Algorithm 2 and consists of the VND, where $x$ is the solution passed as an input parameter.

---

**Algorithm 2:** VND.

1 Let $x$ be an input solution.
2 $k \leftarrow 1$.
3 **while** $k \leq k_{max}$ **do**
4 | $x' \leftarrow$ the best neighbor solution in $N_k(x)$.
5 | **if** $f(x') > f(x)$ **then**
6 | | $x \leftarrow x'$.
7 | | $k \leftarrow 1$.
8 | **else**
9 | | $k \leftarrow k + 1$.

10 **return** $x$.

---

## 3.2 Neighborhood Structures

We consider six neighborhood structures that are used in lines 5 and 4 of Algorithms 1 and 2, respectively. These structures are based on swap, remove, and insertion movements. In particular, they are:

- $N_1$: one doctor, $m_t$, is selected and then two patients, $j_1$ and $j_2$, assigned to this doctor are also selected. Hence, swap $j_1$ for $j_2$ and vice-versa;
- $N_2$: one doctor, $m_t$, is selected and then two patients, $j_1$ and $j_2$, assigned to this doctor are also selected. Hence, insert $j_1$ before $j_2$;
- $N_3$: two doctors, $m_t$ and $m_s$, are selected and then two patients, $j_1$ from $m_t$ and $j_2$ from $m_s$, each one assigned to each doctor, are also selected. Hence, swap $j_1$ for $j_2$ and vice-versa;
- $N_4$: two doctors, $m_t$ and $m_s$, are selected and then two patients, $j_1$ from $m_t$ and $j_2$ from $m_s$, each one assigned to each doctor, are also selected. Hence, remove $j_1$ from $m_t$ and insert it before $j_2$;
- $N_5$: one doctor, $m_t$, is selected and then tree patients, $j_1$, $j_2$ and $j_3$, assigned to this doctor are also selected, where $j_1 < j_2$, $j_3$ is not in the sequence of patients $\{j_1, \ldots, j_2\}$, and the size of this sequence is limited to half of the number of patients assigned to $m_t$. Hence, insert the sequence $\{j_1, \ldots, j_2\}$ before $j_3$;
- $N_6$: two doctors, $m_t$ and $m_s$, are selected and then tree patients, $j_1$ and $j_2$ from $m_t$, and $j_3$ from $m_s$ are also selected, where $j_1 < j_2$ and the sequence of patients $\{j_1, \ldots, j_2\}$ has its size limited to half of the number of patients assigned to $m_t$. Hence, remove the sequence $\{j_1, \ldots, j_2\}$ from $m_t$ and insert it before $j_3$.

### 3.3  Dynamic Version

We firstly propose a greedy heuristic that schedules patients to doctors according to the patients' weight. Each weight is associated with a queue $Q$ that holds the patients organized (i.e., sorted) according to their due date (i.e., the earliest due date first). The first queue, $Q_1$, holds the patients with the highest priority (i.e., the maximum weight), and so on, until the last queue, $Q_l$, which contains the patients with the lowest weight.

The proposed greedy heuristic is described in Algorithm 3. We assume the time horizon starts at $t = 0$, and it continues while patients are arriving. The rule consists of assigning the highest priority patients as soon as possible to the free doctors. Notice that low priority patients may wait indefinitely for free doctors. To avoid this, patients are moved from their current queues to the next ones if their due dates are violated. A new due date is then set for these patients following the priority weights that the queues represent.

---

**Algorithm 3:** DYN1 - Greedy heuristic for the dynamic version.

1  **for** $t \leftarrow 0, 1, \ldots$ **do**
2      Add the patients that arrive at time $t$ into the queues $Q_1, \ldots, Q_l$ according to the patients' priority weight.
3      **foreach** *doctor* $m_f \in M$ *that is free at time* $t$ **do**
4          **for** $q \leftarrow 1, \ldots, l$ **do**
5              **if** $Q_q \neq \emptyset$ **then**
6                  $j \leftarrow$ (remove) the first patient in $Q_q$.
7                  Assign $j$ to $m_f$.
8                  Go to line 3.
9      Update the queues according to the patients' due date.

---

The other heuristic proposed for the dynamic version is in Algorithm 4, taking advantage of the GVNS. For each time $t$ on the horizon, we create a list $L$ with the patients that have just arrived at time $t$, so their release date is equal to $t$. If $L$ is not empty, we optimize the current solution $x$ taking into account the patients of $L$. Given the current schedule at $x$, we fix all the patients whose start time is less than $t$. Notice that there is no preemption after a patient starts, and if the start time of a patient is set after its due date, its tardiness will be greater than zero. We initially schedule the patients of $L$ randomly to $x$ (their start times are always equal to or greater than $t$). Next, the GVNS optimizes all the non-fixed (i.e., non-served yet and that have just arrived) patients in $x$, considering the patients already fixed.

## 4  Computational Experiments

All the algorithms were coded in the C++ programming language. The experiments were carried out in a computer with processor Intel Core i5 2.9 GHz,

---

**Algorithm 4:** DYN2 - GVNS for the dynamic version.

---

1  Let $x$ be an empty solution.
2  **for** $t \leftarrow 0, 1, \ldots$ **do**
3     $L \leftarrow$ list with the patients that have just arrived at time $t$.
4     **if** $L$ *is not empty* **then**
5        Fix all the patients $j$ in $x$ whose $s_j < t$.
6        Assign randomly the patients of $L$ to doctors in $x$.
7        Apply the GVNS (Algorithm 1) to optimize the schedule of all non-fixed patients in $x$.

8  **return** $x$.

---

16 GB of RAM, and macOS Mojave 10.14 as the operating system. The algorithms' parameters consider the number of neighborhood structures, which is set to 6, and the stopping criteria for the outer loop (line 2 of Algorithm 1). We assume a maximum number of iterations and a maximum number of consecutive iterations without improving the best solution for the latter. We defined these parameters after a trial and error process of calibration, achieving the values of 250 and 20, respectively, for the while loop in line 2 of Algorithm 1 (i.e., static problem), and 20 and 5 iterations, respectively, for each call of the GVNS in line 7 of Algorithm 4 (i.e., dynamic problem). We first detail the benchmark instances used to evaluate our algorithms, and then we present and discuss the obtained results.

### 4.1 Benchmark Instances

We generated a set of random instances to evaluate the proposed algorithms. For that, the service times $p_j$ and priority weights $w_j$ of the patients were generated from a uniform distribution in the intervals $[30, 180]$ and $[1, 5]$, respectively. The arrival dates of the patients at the ED were generated based on the scheme used by [17] and [24] to generate release dates for a machine scheduling problem. Following this scheme, the arrival dates $r_j$ were drawn from an uniform distribution in the interval $[0, 50.5 \frac{n}{m} \alpha]$, where $\alpha \in \{0.2, 0.4, 0.6, 0.8, 1.0, 1.25, 1.5, 1.75, 2.0, 3.0\}$ is a parameter that controls how dispersed are the arrival dates. The due dates $d_j$, in turn, depend on $r_j$ and $w_j$ in such a way that $d_j = r_j + 60(w_j - 1)$. Also, for the dynamic case, information about each patient is available when the patients arrive at the ED. We created instances with $n \in \{20, 50, 100, 150, 200\}$ and $m \in \{3, 5\}$. Then, for each combination of $n$, $m$, and $\alpha$, we generated one instance called Nn-Mm-Rz, where $z = 0$ means the first $\alpha$ value (0.2), $z = 1$ means the second $\alpha$ value (0.4), and so on.

### 4.2 Computational Results

Tables 1 and 2 present the results of all instances. Each row of these tables contains the following information: instance name, time in seconds, and solution value (i.e., weighted tardiness) for the static and dynamic problems. In the algorithms for the dynamic problem, we also present the percentage deviation in the solution value when compared to the static problem's solution. Solutions (i.e., the patients' weighted tardiness) whose values are equal to zero are marked in bold.

Observing the results of Table 1, the GVNS obtained 7 solutions equal to zero for the static problem on the instances with 20 patients. This number reduces to 1 with DYN1 and 5 with DYN2. For the instances with 50 patients, the number of solutions equal to zero is 2, for the static problem solved with the GVNS, and 1, for the dynamic problem solved with the DYN2 heuristic, while DYN1 was not able to achieve this. Regarding the percentage deviation between the solutions of the dynamic problem with those of the static problem, the average value is 118.2%, for DYN1, and 54.4%, for DYN2, if considering the instances with 20 patients. For the instances with 50 patients, these numbers are 95.4% and 29.2%, respectively. As the problem is dynamic, we notice that DYN2, which uses the GVNS each time a patient arrives, has performed better than DYN1, which schedules according to queues of priorities. About the computational time, the average value considering all the 40 instances in the table is 4.9 s, for the GVNS in the static problem, 0.1 s, for DYN1, and 2.0 s, for DYN2.

The results of Table 2 refer to the instances with 100 and 150 patients. For the instances with 100 patients, the GVNS obtained 2 solutions with a value equal to zero for the static problem, while DYN1 was not able to do this, and DYN2 did it for 1 instance. For the instances with 200 patients, this was not achieved for any problem (static or dynamic). Considering the percentage deviation between the solutions of DYN1 and DYN2 with the solutions of the GVNS for the static problem, the average values are 89.5% and 19.8%, respectively, for the instances with 100 patients. These values are 85.7% and 18.2% for the instances with 150 patients. Once again, DYN2 has outperformed DYN1 in terms of solution cost. In terms of computational time, the average value considering all the 40 instances in the table is 505.4 s, for the GVNS in the static problem, 0.1 s, for DYN1, and 301.2 s, for DYN2.

When considering all results in Tables 1 and 2, the overall average percentage deviation between the solutions of DYN1 and DYN2 with the solutions of the GVNS for the static problem are 97.2% and 30.4%, respectively. Concerning the computational time, the GVNS for the static problem has an overall average of 255.2 s, while this value is 0.1 s, for DYN1, and 151.6 s, for DYN2. DYN1 is a greedy heuristic that is relatively easy to apply in practice and requires less computational time to run, even for many patients. On the other hand, DYN2 is more sophisticated. It uses the GVNS to optimize the schedule. However, it is more efficient than DYN1 in terms of solutions, with a lower percentage deviation between its solutions and the GVNS for the static problem.

**Table 1.** Results of the instances with 20 and 50 patients.

| Instance | VNS - static | | DYN1 - with queues | | | DYN2 - with VNS | | |
|---|---|---|---|---|---|---|---|---|
| | Time | Sol. | Time (s) | Sol. | Dev. (%) | Time (s) | Sol. | Dev. (%) |
| N20-M3-R0 | 0.2 | 7606 | 0.1 | 14720 | 63.7 | 0.2 | 7624 | 0.2 |
| N20-M3-R1 | 0.4 | 5216 | 0.1 | 8459 | 47.4 | 0.2 | 5236 | 0.4 |
| N20-M3-R2 | 0.2 | 4543 | 0.1 | 8081 | 56.1 | 0.2 | 4545 | 0.0 |
| N20-M3-R3 | 0.3 | 2926 | 0.1 | 5740 | 64.9 | 0.1 | 3516 | 18.3 |
| N20-M3-R4 | 0.5 | 0 | 0.1 | 463 | 200.0 | 0.1 | 77 | 200.0 |
| N20-M3-R5 | 0.3 | 241 | 0.1 | 431 | 56.5 | 0.1 | 873 | 113.5 |
| N20-M3-R6 | 0.5 | 3964 | 0.1 | 8039 | 67.9 | 0.1 | 4239 | 6.7 |
| N20-M3-R7 | 0.3 | 796 | 0.1 | 2177 | 92.9 | 0.1 | 1295 | 47.7 |
| N20-M3-R8 | 0.4 | 52 | 0.1 | 135 | 88.8 | 0.1 | 316 | 143.5 |
| N20-M3-R9 | 0.1 | 0 | 0.1 | 5 | 200.0 | 0.1 | 0 | 0.0 |
| N20-M5-R0 | 0.3 | 769 | 0.1 | 1413 | 59.0 | 0.2 | 796 | 3.5 |
| N20-M5-R1 | 0.2 | 1085 | 0.1 | 2862 | 90.0 | 0.1 | 1182 | 8.6 |
| N20-M5-R2 | 0.3 | 192 | 0.1 | 2178 | 167.6 | 0.1 | 412 | 72.8 |
| N20-M5-R3 | 0.5 | 130 | 0.1 | 438 | 108.5 | 0.1 | 277 | 72.2 |
| N20-M5-R4 | 0.2 | 0 | 0.1 | 122 | 200.0 | 0.1 | 0 | 0.0 |
| N20-M5-R5 | 0.1 | 0 | 0.1 | 140 | 200.0 | 0.1 | 0 | 0.0 |
| N20-M5-R6 | 0.1 | 0 | 0.1 | 120 | 200.0 | 0.1 | 0 | 0.0 |
| N20-M5-R7 | 0.2 | 0 | 0.1 | 235 | 200.0 | 0.1 | 20 | 200.0 |
| N20-M5-R8 | 0.1 | 0 | 0.1 | 200 | 200.0 | 0.1 | 78 | 200.0 |
| N20-M5-R9 | 0.1 | 0 | 0.1 | 0 | 0.0 | 0.1 | 0 | 0.0 |
| **Average** | 0.3 | − | 0.1 | − | 118.2 | 0.1 | − | 54.4 |
| N50-M3-R0 | 13.4 | 48190 | 0.1 | 89687 | 60.2 | 13.3 | 48495 | 0.6 |
| N50-M3-R1 | 22.0 | 49556 | 0.1 | 98229 | 65.9 | 13.2 | 49784 | 0.5 |
| N50-M3-R2 | 16.6 | 39662 | 0.1 | 73301 | 59.6 | 8.1 | 39994 | 0.8 |
| N50-M3-R3 | 14.0 | 26157 | 0.1 | 64560 | 84.7 | 6.2 | 26819 | 2.5 |
| N50-M3-R4 | 12.0 | 37305 | 0.1 | 75064 | 67.2 | 6.0 | 41144 | 9.8 |
| N50-M3-R5 | 11.2 | 19506 | 0.1 | 47331 | 83.3 | 1.7 | 19506 | 0.0 |
| N50-M3-R6 | 13.8 | 7657 | 0.1 | 23996 | 103.2 | 0.9 | 9343 | 19.8 |
| N50-M3-R7 | 8.3 | 1594 | 0.1 | 4335 | 92.5 | 0.4 | 3118 | 64.7 |
| N50-M3-R8 | 7.0 | 252 | 0.1 | 2126 | 157.6 | 0.1 | 1016 | 120.5 |
| N50-M3-R9 | 3.8 | 0 | 0.1 | 175 | 200.0 | 0.1 | 0 | 0.0 |
| N50-M5-R0 | 7.1 | 27692 | 0.1 | 44234 | 46.0 | 7.7 | 28258 | 2.0 |
| N50-M5-R1 | 5.0 | 21486 | 0.1 | 44119 | 69.0 | 5.2 | 22495 | 4.6 |
| N50-M5-R2 | 5.7 | 14482 | 0.1 | 33754 | 79.9 | 5.0 | 15004 | 3.5 |
| N50-M5-R3 | 10.1 | 12517 | 0.1 | 32416 | 88.6 | 4.6 | 13362 | 6.5 |
| N50-M5-R4 | 7.4 | 13769 | 0.1 | 28113 | 68.5 | 2.2 | 14572 | 5.7 |
| N50-M5-R5 | 10.7 | 7394 | 0.1 | 18988 | 87.9 | 1.2 | 9115 | 20.8 |
| N50-M5-R6 | 8.4 | 2764 | 0.1 | 6182 | 76.4 | 0.9 | 5136 | 60.1 |
| N50-M5-R7 | 6.0 | 7900 | 0.1 | 18803 | 81.7 | 0.9 | 9525 | 18.7 |
| N50-M5-R8 | 6.9 | 264 | 0.1 | 1389 | 136.1 | 0.2 | 405 | 42.2 |
| N50-M5-R9 | 2.9 | 0 | 0.1 | 224 | 200.0 | 0.1 | 185 | 200.0 |
| **Average** | 9.6 | − | 0.1 | − | 95.4 | 3.9 | − | 29.2 |

**Table 2.** Results of the instances with 100 and 150 patients.

| Instance | VNS - static | | DYN1 - with queues | | | DYN2 - with VNS | | |
|---|---|---|---|---|---|---|---|---|
| | Time | Sol. | Time (s) | Sol. | Diff. (%) | Time (s) | Sol. | Diff. (%) |
| N100-M3-R0 | 248.4 | 263437 | 0.1 | 434323 | 49.0 | 352.9 | 265621 | 0.8 |
| N100-M3-R1 | 243.6 | 229351 | 0.1 | 421520 | 59.0 | 219.1 | 229351 | 0.0 |
| N100-M3-R2 | 225.3 | 146126 | 0.1 | 300384 | 69.1 | 170.9 | 147851 | 1.2 |
| N100-M3-R3 | 171.9 | 139286 | 0.1 | 260380 | 60.6 | 114.9 | 141554 | 1.6 |
| N100-M3-R4 | 90.5 | 117357 | 0.1 | 264841 | 77.2 | 93.0 | 119129 | 1.5 |
| N100-M3-R5 | 163.5 | 65989 | 0.1 | 182378 | 93.7 | 33.0 | 70582 | 6.7 |
| N100-M3-R6 | 281.9 | 60700 | 0.1 | 123764 | 68.4 | 24.3 | 67205 | 10.2 |
| N100-M3-R7 | 189.3 | 23643 | 0.1 | 55736 | 80.9 | 4.6 | 24455 | 3.4 |
| N100-M3-R8 | 239.8 | 22923 | 0.1 | 62755 | 93.0 | 3.4 | 26725 | 15.3 |
| N100-M3-R9 | 96.1 | 0 | 0.1 | 799 | 200.0 | 0.1 | 68 | 200.0 |
| N100-M5-R0 | 132.9 | 120418 | 0.1 | 210010 | 54.2 | 180.8 | 120918 | 0.4 |
| N100-M5-R1 | 128.0 | 113361 | 0.1 | 197608 | 54.2 | 118.9 | 114741 | 1.2 |
| N100-M5-R2 | 88.0 | 100983 | 0.1 | 205304 | 68.1 | 136.1 | 101509 | 0.5 |
| N100-M5-R3 | 78.1 | 78868 | 0.1 | 155045 | 65.1 | 68.7 | 82185 | 4.1 |
| N100-M5-R4 | 133.9 | 78029 | 0.1 | 158456 | 68.0 | 67.7 | 78170 | 0.2 |
| N100-M5-R5 | 261.5 | 25251 | 0.1 | 79036 | 103.1 | 37.4 | 29274 | 14.8 |
| N100-M5-R6 | 115.7 | 23463 | 0.1 | 68326 | 97.8 | 11.1 | 30789 | 27.0 |
| N100-M5-R7 | 159.0 | 6273 | 0.1 | 23999 | 117.1 | 4.2 | 10467 | 50.1 |
| N100-M5-R8 | 192.9 | 4480 | 0.1 | 15901 | 112.1 | 2.7 | 8062 | 57.1 |
| N100-M5-R9 | 26.9 | 0 | 0.1 | 145 | 200.0 | 0.2 | 0 | 0.0 |
| **Average** | 163.3 | – | 0.1 | – | 89.5 | 82.2 | – | 19.8 |
| N150-M3-R0 | 1061.5 | 648760 | 0.1 | 1067988 | 48.8 | 2396.9 | 650043 | 0.2 |
| N150-M3-R1 | 1756.3 | 483268 | 0.1 | 941882 | 64.4 | 1901.9 | 483268 | 0.0 |
| N150-M3-R2 | 576.4 | 432851 | 0.1 | 813579 | 61.1 | 1313.5 | 432851 | 0.0 |
| N150-M3-R3 | 779.6 | 287406 | 0.1 | 611584 | 72.1 | 706.1 | 287406 | 0.0 |
| N150-M3-R4 | 1025.2 | 257484 | 0.1 | 542107 | 71.2 | 584.6 | 267243 | 3.7 |
| N150-M3-R5 | 1920.2 | 174018 | 0.1 | 452553 | 88.9 | 217.6 | 174018 | 0.0 |
| N150-M3-R6 | 893.0 | 88666 | 0.1 | 221176 | 85.5 | 88.7 | 98642 | 10.7 |
| N150-M3-R7 | 959.3 | 61445 | 0.1 | 214027 | 110.8 | 34.0 | 67316 | 9.1 |
| N150-M3-R8 | 1234.9 | 36568 | 0.1 | 106471 | 97.7 | 10.9 | 42389 | 14.7 |
| N150-M3-R9 | 596.0 | 487 | 0.1 | 3351 | 149.2 | 0.3 | 1617 | 107.4 |
| N150-M5-R0 | 977.3 | 302481 | 0.1 | 574030 | 62.0 | 904.9 | 304115 | 0.5 |
| N150-M5-R1 | 451.1 | 257798 | 0.1 | 476820 | 59.6 | 864.9 | 262776 | 1.9 |
| N150-M5-R2 | 738.1 | 212477 | 0.1 | 440777 | 69.9 | 595.9 | 212477 | 0.0 |
| N150-M5-R3 | 945.3 | 181907 | 0.1 | 418622 | 78.8 | 377.0 | 185250 | 1.8 |
| N150-M5-R4 | 641.6 | 134576 | 0.1 | 324238 | 82.7 | 249.4 | 144552 | 7.1 |
| N150-M5-R5 | 1135.8 | 80342 | 0.1 | 204213 | 87.1 | 78.3 | 84210 | 4.7 |
| N150-M5-R6 | 377.8 | 64906 | 0.1 | 175178 | 91.9 | 55.4 | 71206 | 9.3 |
| N150-M5-R7 | 318.8 | 32488 | 0.1 | 109132 | 108.2 | 20.5 | 39969 | 20.6 |
| N150-M5-R8 | 315.7 | 7686 | 0.1 | 20016 | 89.0 | 3.2 | 14167 | 59.3 |
| N150-M5-R9 | 246.3 | 175 | 0.1 | 897 | 134.7 | 1.0 | 634 | 113.5 |
| **Average** | 847.5 | – | 0.1 | – | 85.7 | 520.2 | – | 18.2 |

# 5   Concluding Remarks

The scheduling of patients to doctors in emergency departments is a dynamic problem that requires fast and accurate decisions while minimizing the weighted tardiness related to the waiting time of patients. We proposed a greedy heuristic based on weighted queues in which the patients' weight is increased when they wait beyond their due date. We also proposed a heuristic that uses a general variable neighborhood search to optimize the schedule. The GVNS has the variable neighborhood descent as the local search, where six structures based on swap, remove, and insertion movements are defined. The GVNS is also applied to solve the static problem and then provide a reference solution for the dynamic problem.

Results of 80 instances have indicated that the best solutions are achieved if information about the problem is known in advance (i.e., the static problem). As this is not possible in real situations, the dynamic problem is better solved when the GVNS is used instead of only a greedy heuristic based on weighted queues. While in the static problem, the number of solutions with zero tardiness is equal to 12, in the dynamic problem, this number reduces to 1, with the greedy heuristic, and 7, when the GVNS is considered. In terms of computational time, the greedy heuristic is much less expensive; however, its solutions are far from those of the static problem. As an emergency department should have to be concerned with the patients' satisfaction and health, a method that performs better should be preferable, even if it requires more computational power to run.

Future works will consider a mathematical model for the static version (see, e.g., [16]). Besides that, other characteristics the problem may assume will be investigated, for example, to reveal some information of patients before the arrival to the emergency department. We are also interested in exploring other neighborhood structures and local searches in the VNS framework.

**Acknowledgments.** This research was partially funded by Health and Medical Research Fund, Food and Health Bureau, the Hong Kong SAR Government (grant 14151771), the Early Career Scheme (ECS), Research Grants Council (RGC) of Hong Kong (grant 27200419), the National Council for Scientific and Technological Development (CNPq - grants 234814/2014-4 and 308312/2016-3), the State of Goiás Research Foundation (FAPEG), and the University of Modena and Reggio Emilia (grant FAR 2018 - Analysis and optimization of health-care and pharmaceutical logistic processes).

# References

1. Baptiste, P., Jouglet, A., Savourey, D.: Lower bounds for parallel machine scheduling problems. Int. J. Oper. Res. **3**(6), 643–664 (2008)
2. Bulhões, T., Sadykov, R., Subramanian, A., Uchoa, E.: On the exact solution of a large class of parallel machine scheduling problems. J. Sched. **23**(4), 411–429 (2020). https://doi.org/10.1007/s10951-020-00640-z

3. Burke, E.K., Li, J., Qu, R.: A hybrid model of integer programming and variable neighbourhood search for highly-constrained nurse rostering problems. Eur. J. Oper. Res. **203**(2), 484–493 (2010)
4. Dellaert, N., Jeunet, J.: A variable neighborhood search algorithm for the surgery tactical planning problem. Comput. Oper. Res. **84**, 216–225 (2017)
5. Derlet, R.W., Richards, J.R.: Overcrowding in the nation's emergency departments: complex causes and disturbing effects. Ann. Emerg. Med. **35**(1), 63–68 (2000)
6. Di Somma, S., Paladino, L., Vaughan, L., Lalle, I., Magrini, L., Magnanti, M.: Overcrowding in emergency department: an international issue. Intern. Emerg. Med. **10**(2), 171–175 (2014). https://doi.org/10.1007/s11739-014-1154-8
7. Ding, Y., Park, E., Nagarajan, M., Grafstein, E.: Patient prioritization in emergency department triage systems: an empirical study of the canadian triage and acuity scale (CTAS). Manuf. Serv. Oper. Manag. **21**(4), 723–741 (2019)
8. Fernandes, C.M., et al.: Five-level triage: a report from the ACEP/ENA five-level triage task force. J. Emerg. Nurs. **31**(1), 39–50 (2005)
9. Fleszar, K., Hindi, K.S.: Solving the resource-constrained project scheduling problem by a variable neighbourhood search. Eur. J. Oper. Res. **155**(2), 402–413 (2004)
10. Frifita, S., Masmoudi, M., Euchi, J.: General variable neighborhood search for home healthcare routing and scheduling problem with time windows and synchronized visits. Electron. Notes Discret. Math. **58**, 63–70 (2017)
11. Graham, R., Lawler, E., Lenstra, J., Kan, A.: Optimization and approximation in deterministic sequencing and scheduling: a survey. In: Hammer, P.L., Johnson, E.L., Korte, B. (eds.) Discrete Optimization II Proceedings of the Advanced Research Institute on Discrete Optimization and Systems Applications, Annals of Discrete Mathematics, vol. 5, pp. 287–326 (1979)
12. Green, L.V., Soares, J., Giglio, J.F., Green, R.A.: Using queueing theory to increase the effectiveness of emergency department provider staffing. Acad. Emerg. Med. **13**(1), 61–68 (2006)
13. Hansen, P., Mladenović, N.: Variable neighborhood search. In: Burke, E.K., Kendall, G. (eds.) Search Methodologies. Springer, Boston (2005). https://doi.org/10.1007/0-387-28356-0_8
14. Jouglet, A., Savourey, D.: Dominance rules for the parallel machine total weighted tardiness scheduling problem with release dates. Comput. Oper. Res. **38**(9), 1259–1266 (2011)
15. Kramer, A., Subramanian, A.: A unified heuristic and an annotated bibliography for a large class of earliness-tardiness scheduling problems. J. Sched. **22**, 21–57 (2019)
16. Kramer, A., Dell'Amico, M., Iori, M.: Enhanced arc-flow formulations to minimize weighted completion time on identical parallel machines. Eur. J. Oper. Res. **275**(1), 67–79 (2019)
17. Kramer, A., Dell'Amico, M., Feillet, D., Iori, M.: Scheduling jobs with release dates on identical parallel machines by minimizing the total weighted completion time. Comput. Oper. Res. **123**, 105018 (2020)
18. Kuo, Y.H.: Integrating simulation with simulated annealing for scheduling physicians in an understaffed emergency department. HKIE Trans. **21**(4), 253–261 (2014)
19. Kuo, Y.H., Leung, J.M., Graham, C.A., Tsoi, K.K., Meng, H.M.: Using simulation to assess the impacts of the adoption of a fast-track system for hospital emergency services. J. Adv. Mech. Des. Syst. Manuf. **12**(3), 17-00637:1–17-00637:11 (2018)

20. Kuo, Y.H., Rado, O., Lupia, B., Leung, J.M.Y., Graham, C.A.: Improving the efficiency of a hospital emergency department: a simulation study with indirectly imputed service-time distributions. Flex. Serv. Manuf. J. **2**, 120–147 (2014). https://doi.org/10.1007/s10696-014-9198-7
21. Lenstra, J., Kan, A.R., Brucker, P.: Complexity of machine scheduling problems. In: Hammer, P., Johnson, E., Korte, B., Nemhauser, G. (eds.) Studies in Integer Programming, Annals of Discrete Mathematics, vol. 1, pp. 343–362. Elsevier (1977)
22. Li, J., Pan, Q., Wang, F.: A hybrid variable neighborhood search for solving the hybrid flow shop scheduling problem. Appl. Soft Comput. **24**, 63–77 (2014)
23. Mladenović, N., Hansen, P.: Variable neighborhood search. Comput. Oper. Res. **24**(11), 1097–1100 (1997)
24. Nessah, R., Yalaoui, F., Chu, C.: A branch-and-bound algorithm to minimize total weighted completion time on identical parallel machines with job release dates. Comput. Oper. Res. **35**(4), 1176–1190 (2008)
25. Queiroz, T.A., Mundim, L.R.: Multiobjective pseudo-variable neighborhood descent for a bicriteria parallel machine scheduling problem with setup time. Int. Trans. Oper. Res. **27**(3), 1478–1500 (2020)
26. Richardson, D.B.: Increase in patient mortality at 10 days associated with emergency department overcrowding. Med. J. Aust. **184**(5), 213–216 (2006)
27. Saghafian, S., Hopp, W.J., Van Oyen, M.P., Desmond, J.S., Kronick, S.L.: Complexity-augmented triage: a tool for improving patient safety and operational efficiency. Manuf. Serv. Oper. Manag. **16**(3), 329–345 (2014)
28. Shih, F.Y., et al.: ED overcrowding in Taiwan: facts and strategies. Am. J. Emerg. Med. **17**(2), 198–202 (1999)
29. Wuerz, R.C., Milne, L.W., Eitel, D.R., Travers, D., Gilboy, N.: Reliability and validity of a new five-level triage instrument. Acad. Emerg. Med. **7**(3), 236–242 (2000)

# A GRASP/VND Heuristic for the Heterogeneous Fleet Vehicle Routing Problem with Time Windows

Lucía Barrero, Franco Robledo, Pablo Romero<sup>(✉)</sup>, and Rodrigo Viera

Instituto de Computación, INCO,
Facultad de Ingeniería - Universidad de la República, Montevideo, Uruguay
{lucia.barrero,frobledo,promero,rodrigo.viera}@fing.edu.uy

**Abstract.** The Heterogeneous Fleet Vehicle Routing Problem with Time Windows (HFVRPTW) is here introduced. This combinatorial optimization problem is an extension of the well-known Vehicle Routing Problem (VRP), which belongs to the $\mathcal{NP}$-Hard class. As a corollary, our problem belongs to this class, a fact that promotes the development of approximative methods.

A mathematical programming formulation for the HFVRPTW is presented, and an exact solution method using CPLEX is implemented. A GRASP/VND methodology is also developed, combining five different local searches. The effectiveness of our proposal is studied in relation with the exact solver. Our proposal outperforms the exact CPLEX in terms of CPU times, and finds even better solutions under large-sized instances, where the exact solver halts after ten hours of continuous execution.

**Keywords:** Combinatorial optimization problem · Vehicle routing problem · HFVRPTW · Computational complexity · GRASP · VND

## 1 Motivation

The transport industry employs more than 10 million people and it represents roughly the 5% of the Gross Domestic Product (GDP) of the European Union. Furthermore, logistics such as transport and storage account for 10%–15% of the cost of a finished product. In practice, this means that even a small relative reduction in the cost of logistics and transportation means huge savings.

Usually, large-scale corporations in the transport sector are mostly dedicated to savings, and an efficient delivery of goods and services. However, transport also represents an important source of $CO_2$ emissions, and traffic congestion. In synthesis, a smart vehicle routing engineering is not only meaningful in terms of savings, but also implies a responsible care of the environment.

Operational researchers are engaged with society, and try their best to develop mathematical models that are suitable for realistic transportation problems. A celebrated combinatorial problem is known as the Traveling Salesman

© Springer Nature Switzerland AG 2021
N. Mladenovic et al. (Eds.): ICVNS 2021, LNCS 12559, pp. 152–165, 2021.
https://doi.org/10.1007/978-3-030-69625-2_12

Problem, or TSP. We are given non-negative costs in the edges of a complete network, and the goal is to find the cheapest Hamiltonian tour (i.e., visiting all the nodes of the network). The decision version for the TSP belongs to the class of $\mathcal{NP}$-Complete problems, and it is included in Karp list [8]. A natural generalization is the Vehicle Routing Problem, or VRP. In the VRP, we are given a fleet of vehicles, and we should determine the optimal set of routes in order to serve a given number of customers, starting and ending at the depot. The reader can appreciate that the TSP is a special VRP with a single vehicle; thus, the VRP belongs to the $\mathcal{NP}$-Hard class. Given its paramount importance, several variations in the basic VRP model appear in the literature, adding time-windows for customer delivery, heterogeneous fleets, one-way or two-way routes, dynamic demands, among many others. The reader can consult the recent survey for the different variants of the VRP and its applicability to different contexts [9].

To the best of our knowledge, there is no model that simultaneously combines heterogeneous fleets and time-windows, with a penalty factor due to overtime. The contributions of this paper can be summarized in the following items:

1. The Heterogeneous Fleet Vehicle Routing Problem with Time Windows (HFVRPTW) is introduced.
2. We formally prove that the HFVRPTW belongs to the $\mathcal{NP}$-Hard class.
3. As a consequence, a GRASP/VND methodology is proposed.
4. A novel mathematical programming formulation for the HFVRPTW is presented. It represents an adaptation of the previous formulation proposed in [7], adding a penalty due to overtime.
5. The effectiveness of our proposal with respect to an exact solution implemented in CPLEX is studied. The activity of the different local searches of our GRASP/VND methodology is also studied.

The document is organized in the following manner. The related work is presented in Sect. 2. A formal description for the HFVRPTW is presented in Sect. 3; its $\mathcal{NP}$-Hardness is also established. A GRASP/VND solution is introduced in Sect. 4. Numerical results are presented in Sect. 5. Section 6 contains concluding remarks and trends for future work.

## 2   Related Work

The classical VRP is presented by Dantzig as a generalization of the TSP [4]. The problem is there motivated by fuel distribution, trying to find the optimal routing of a fleet between a depot and several stations. In general, the VRP consists of how to share customers geographically distributed by a given fleet of vehicles, based on one or multiple depots. The goal is to fulfill the customer demands, finding adequate routes starting and ending at the depot. Rapidly, the VRP found an impressive diversity of applications, ranging from transport network design to efficient garbage collectors. Current VRP models include more realistic assumptions (such as traffic congestion and time-windows for the customers), given the greater possibilities in processing resources. In [1], Baldacci presents a framework for exact algorithms useful for several variations of the VRP, such as capacitated VRP, VRP with Time Windows (VRPTW), pick-up and delivery, multi-depot VRP, among others. In the Heterogeneous Fleet VRP, we are given vehicles with different capacities, and the goal is to design a minimum cost solution meeting the customer demands, starting and ending at the central depot. A fixed cost is associated to the vehicle-type, while a variable cost is proportional to the distance of the tours.

An exact Branch and Cut solution for the HFVRP is proposed in [11], adapting the most competitive exact algorithms for the problem such as route enumeration and extended capacity cuts for large-sized instances.

Other works address the VRP with Time-Windows (VRPTW), where the TW have either soft or hard constraints. In the hard constraint, an early vehicle can wait until the customer is available. In a soft constraint, a penalty is carried to the objective when the constraint is not satisfied. Historical works for the soft VRPTW show that an incorrect usage of a Tabu Search the TW can have a negative impact in the cost [10,14,17]. A hybrid solution for the VRPTW is proposed in [16], that jointly considers Large Neighborhood Search (LNS) and a Bat Algorithm (BA), inspired by the eco-location of bats. The results were satisfactory, under benchmarks with 100 customers.

In [2], a two-phase solution combines a Construction phase with Tabu Search, to avoid locally optimum solutions. The solution reduce the distances, in a practical industrial application. A hierarchic *cluster-first route-second* solution for a large super-market chain is proposed in [3], with remarkable benefits with respect to a naive solution.

In this work, we combine Heterogeneous Fleet with a new concept of soft constraint with overtime. Our formulation is adapted from the mathematical programming presented in [7]. The reader is invited to consult the recent review on the VRP for other variations of this problem [9].

## 3    Problem and Complexity

In this section, a formal combinatorial optimization problem is introduced. The hardness of the problem is also established.

### 3.1    Formulation

The exact formulation is based on the integer linear programming model defined in [7]. However, we consider flexible time-windows instead, where delays are penalized with a cost (i.e., an additive term in the objective function). Consider a complete graph $G = (V, E)$ where:

- $V = \{0, 1, \dots, n\}$, being 0 the depot and $N = \{1, \dots, n\}$ the customers.
- $E = \{(i, j) : 0 \leq i, j \leq n, i \neq j\}$ represent the links between the nodes.
- $t_{ij}$ is the required time to cross the link $(i, j)$.

All the customers must be visited, and the following information is known for each customer $i \in N$:

- $d_i$ is a fixed demand for customer $i$.
- $s_i$ represents the required time for a vehicle to service the customer $i$.
- $[e_i, l_i]$ is the time-window (available and deadline) for customer $i$. This window is not a hard constraint (a penalty occurs if it is not respected).
- $ot_i$ is the *overtime*, or the tolerance after the deadline. It is found with the following expression: $ot_i = \omega(l_i - e_i)$ for some known factor $\omega : 0 \leq \omega \leq 1$. The extended Time Window (TW) is then $[e_i, l_i + ot_i]$. A penalty occurs if the vehicle meets customer $i$ during the interval $[l_i, l_i + ot_i]$.

With respect to the depot, we know that:

- $[e_0, l_0] = [E, L]$ is the time-window for the depot.
- $d_0 = s_0 = 0$, since in the depot there is no demand nor service.

The fleet is modeled as $K = \{1, ..., k\}$, $C$ represents the vehicle-types, and $S_c$ the set of $c$-type vehicles. For each vehicle, we are given:

- $q_c$ is the capacity.
- $f_c$ is its fixed-cost.
- $\alpha_c$ is its variable-cost.
- $n_c$ is the number of available type-$c$ vehicles.

We consider the following set of decision variables:

- $x_{ij}^k = 1$ iff the vehicle $k$ visits the link $(i, j)$; 0 otherwise.
- $a_{ik}$: time which the vehicle $k$ reaches the customer $i$.
- $o_{ik}$: overtime of vehicle $k$ for the customer $i$.

We also assume that the following parameters are known:

- $M = \max\limits_{(i,j \in V)} (l_i + ot_i + t_{ij} + s_i - e_j)$: represents the longest time consumed between any two customers.
- $\rho$: represents the penalty associated to overtime.

The HFVRPTW can be formulated as follows:

$$\min \sum_{c \in C} f_c \sum_{k \in S_c} \sum_{j \in N} x_{0j}^k + \sum_{c \in C} \alpha_c \sum_{k \in S_c} \sum_{\substack{i,j \in V, \\ i \neq j}} t_{ij} x_{ij}^k + \sum_{k \in K, i \in N} o_{ik} * \rho \quad (1)$$

s.t.:

$$\sum_{k \in K} \sum_{\substack{j \in V, \\ i \neq j}} x_{ij}^k = 1 \ \forall i \in N \quad (2)$$

$$\sum_{j \in N} x_{0j}^k \leq 1 \ \forall k \in K \quad (3)$$

$$\sum_{i \in N} x_{i0}^k \leq 1 \ \forall k \in K \quad (4)$$

$$\sum_{i \in V} x_{ij}^k = \sum_{i \in V} x_{ji}^k \; \forall j \in V, \; k \in K \tag{5}$$

$$\sum_{i \in N} d_i \sum_{\substack{j \in V, \\ i \neq j}} x_{ij}^k \leq q_c \; \forall k \in S_c, \; c \in C \tag{6}$$

$$a_{ik} + s_i + t_{ij} - M(1 - x_{ij}^k) \leq a_{jk} \; \forall k \in K, \; i \in N, \; j \in V, \; i \neq j \tag{7}$$

$$t_{0i} * x_{0i}^k \leq a_{ik} \; \forall k \in K, \; i \in N \tag{8}$$

$$a_{ik} \leq (l_i + ot_i) \sum_{\substack{j \in V, \\ i \neq j}} x_{ij}^k \; \forall k \in K, \; i \in N \tag{9}$$

$$e_i \sum_{\substack{j \in V, \\ i \neq j}} x_{ij}^k \leq a_{ik} \leq (l_i + ot_i) \sum_{\substack{j \in V, \\ i \neq j}} x_{ij}^k \; \forall k \in K, \; i \in N \tag{10}$$

$$E \leq a_{0k} \leq L + ot_0 \; \forall k \in K \tag{11}$$

$$\sum_{k \in S_c} \sum_{j \in N} x_{0j}^k \leq n_c \; \forall c \in C \tag{12}$$

$$o_{ik} \geq \max(0, a_{ik} - l_i) \geq 0 \; \forall k \in K, \; i \in V \tag{13}$$

$$a_{ik} \geq 0 \; \forall k \in K, \; i \in N \tag{14}$$

$$x_{ij}^k \in \{0, 1\} \; \forall k \in K, \; (i, j) \in E \tag{15}$$

The objective function 1 is an additive cost, considering fixed and variable costs in the vehicles, as well as penalties related to overtime. Constraints 2 state that all the customers must be visited by only one vehicle. The set of Constraints 3, 4 and 5 represent flow conservation, and state that all the vehicles start and end at the depot. Constraints 6 state that the customer demands cannot exceed the capacities of the vehicles. Constraints 7 state the precedence relation between the arrival times of the vehicles to the customers.

Constraints 8 state the first arrival time to the first node in the route. The set of Constraints 9, 10 and 11 model the time-windows for both the customers and the depot, while Constraints 12 bounds the number of available vehicles for each type. Finally, the set of Constraints 13, 14 and 15 define the domain of the respective decision variables.

## 3.2    Hardness

The hardness of the corresponding decision version for the HFVRPTW is straight from the $\mathcal{NP}$-Completeness of Hamiltonian Tour. Recall that a graph $G$ is *Hamiltonian* if there exists an elementary cycle $\mathcal{C} \subseteq G$ that contains all the nodes.

**Definition 1 (Hamiltonian Tour).** *Given a simple graph $G = (V, E)$. Is $G$ Hamiltonian?*

It is known that Hamiltonian Tour belongs to the class of $\mathcal{NP}$-Complete decision problems [5,8].

**Proposition 1.** *The HFVRPTW belongs to the $\mathcal{NP}$-Hard class.*

*Proof.* By reduction from Hamiltonian Tour. Consider an arbitrary graph $G = (V, E)$. We will see that there exists a feasible solution for the HFVRPTW whose cost is not greater than $n = |V|$ if and only if there exists a Hamiltonian tour for $G$.

Consider an instance of HFVRPTW with the complete graph $K_n$ as a ground graph, where $n = |V|$, a single vehicle with cost $\alpha = 1$ and sufficient capacity $q_c = n$ rooted at some arbitrary depot $v \in V$, no penalties and customers with infinite patience. The time to traverse the links $(i, j) \in E$ is always $t_{i,j} = 1$, but $t_{i,j} = n$ if $(i, j) \notin E$. A feasible solution must be a Hamiltonian tour, and its cost is not greater than $n$ if and only if it is strictly included in $G = (V, E)$. Therefore, the HFVRPTW is at least as hard as Hamiltonian Tour.  ∎

Recall that Hamilton Tour is strongly $\mathcal{NP}$-Hard. Thus, the HFVRPTW is hard in the strong sense, and there is no FPTAS for our problem, unless $\mathcal{P} = \mathcal{NP}$.

# 4 Solution

GRASP and VND are well known metaheuristics that have been successfully used to solve many hard combinatorial optimization problems [13]. GRASP is a powerful multi-start process which operates into two phases. A feasible solution is built in a first phase, whose neighborhood is then explored in the Local Search Phase. The second phase is usually enriched by means of different variable neighborhood structures. For instance, VND explores several neighborhood structures in a deterministic order. Its success is based on the simple fact that different neighborhood structures do not usually have the same local minimum. Thus, the resulting solution is simultaneously a locally optimum solution under all the neighborhood structures. The reader is invited to consult the comprehensive Handbook of Heuristics for further information [6]. Here, we develop a GRASP/VND methodology. The main building-blocks of our *Main* algorithm are presented in Fig. 1. An arbitrary input instance $I = (G, t_{ij}, d_i, s_i, e_i, l_i, ot_i, \omega, K, q_c, f_c, \alpha_c, n_c)$ for the HFVRPTW is considered, where the symbols represent the aforementioned variables in the problem formulation. Observe that the whole GRASP/VND solution is executed *iter* times, and the best solution is returned. The parameter $\alpha \in [0, 1]$ trades greediness for randomization during the *Construction* phase, by means of a Restricted Candidate List (RCL). The VND is composed by five local searches, to know, *FleetOpt, Exchange, Relocate, 2 − opt* and *3 − opt*, in the respective order. In the following paragraphs, we describe the Construction phase, as well as the local searches.

---

**Algorithm 1** $sol = Main(I, iter, \alpha)$

---

1: $i \leftarrow 0; sol \leftarrow \emptyset$
2: **while** $i < iter$ **do**
3:     $sol \leftarrow Construction(I, \alpha)$
4:     $\overline{sol} \leftarrow VND(sol, I, FleetOpt, Exchange, Relocate, 2 - opt, 3 - opt)$
5:     **if** $cost(\overline{sol}) < cost(sol)$ **then**
6:         $sol \leftarrow \overline{sol}$
7:     **end if**
8: **end while**
9: **return** $sol$

---

**Fig. 1.** Pseudocode for the $Main$ algorithm.

## 4.1  Construction Phase

Figure 2 presents a full pseudocode for the $Construction$ phase. The following functions are considered:

– $GetClients(data)$: returns the clients in a list for a given dataset.
– $SelectVehicles(vehicles)$: returns a vehicle that is available, and updates the number of available vehicles.
– $GetCapacity(vehicle)$: returns the capacity of a given vehicle.
– $CreateRoute(vehicle, path)$: creates a route using the given $path$. This route is performed with the given $vehicle$.
– $IsFeasible(route, client)$: determines whether it is feasible or not to append the given $client$ at the end of the given $route$, or not.

We need to select vehicles and routes for them, in order to build feasible solutions. We collect all the customers that were not yet visited in the variable $clients$. A metric is considered to decide the priority for the different vehicles. The route is then constructed, that starts and ends at the depot, for that vehicle. A Restricted Candidate List (RCL) is built in order to include different customers in the route, always picking customers from the collection of nonvisited customers in order to meet feasibility. The marginal cost to include some customer is found using the following expression:

$$incr = VariableCost \times t + overtime \times penalty + arrival,$$

being $arrival$ the arrival time at the new candidate. Observe that $incr$ represents an estimation for the marginal increase in the objective, since we need to adjust all the time-windows for the other customers. Nevertheless, the marginal costs are useful to build the RCL, following a classical implementation. We find the least and the greatest marginal costs $i_{min}$ and $i_{max}$, and the RCL consists of the candidates $e$ such that $incr(e) \leq i_{min} + \alpha \times (i_{max} - i_{min})$, being $\alpha \in [0, 1]$ the GRASP parameter that trades greediness for randomization. Finally, a random member belonging to the RCL is inserted into the partial route, and the whole collection of non-visited customers are updated accordingly, with a new

**Algorithm 2** $sol = Construction(instance, vehicles)$

---

1:  $sol \leftarrow \phi$
2:  $clients \leftarrow$ GetClients($instance$)
3:  $newRoute \leftarrow$ **true**
4:  **while** $clients \neq \phi$ **do**
5:      $candidates \leftarrow \phi$
6:      **if** $newRoute$ **then**
7:          $path \leftarrow \{depot\}$
8:          $vehicle \leftarrow$ SelectVehicle($vehicles$)
9:          $q \leftarrow$ GetCapacity($vehicle$)
10:         $route \leftarrow$ CreateRoute($vehicles, path$)
11:         $newRoute \leftarrow$ **false**
12:     **end if**
13:     **for** $client \in clients$ **do**
14:         **if** IsFeasible($route, client$) **then**
15:             $candidates \leftarrow candidates \cup \{client\}$
16:         **end if**
17:     **end for**
18:     **if** $candidates \neq \phi$ **then**
19:         $incr(e) \ \forall \ e \in candidates$
20:         $i_{min} \leftarrow min\{incr(e) : e \in candidates\}$
21:         $i_{max} \leftarrow max\{incr(e) : e \in candidates\}$
22:         $RCL \leftarrow \{e \in candidates : incr(e) \leq i_{min} + \alpha(i_{max} - i_{min})\}$
23:         $client \leftarrow Random(RCL)$
24:         $path \leftarrow path \cup \{client\}$
25:         $q = q - $ GetDemand($client$)
26:         $clients \leftarrow clients \backslash \{client\}$
27:     **end if**
28:     **if** $candidates = \phi \vee q = 0$ **then**
29:         $path \leftarrow path \cup \{depot\}$
30:         $sol \leftarrow sol \cup \{route\}$
31:         $newRoute \leftarrow$ **true**
32:     **end if**
33: **end while**
34: **return** $sol$

---

Fig. 2. Construction phase

evaluation of marginal costs. The route is closed whenever the vehicle capacity is reached, or when there are no more candidates to be included. In that case, the depot node is included.

### 4.2 Local Search Phase - $VND$

Five local searches are called in order, after the *Construction* phase:

1. $Fleet - opt$
2. $Exchange$
3. $Relocate$

4. $2 - opt$
5. $3 - opt$

We followed a strict time-complexity order of the local searches, as suggested in [12]. For practical reasons, we assume that there are more customers than vehicle-types.

**Definition 2 (Fleet-Opt).** *The goal is to change the vehicles. There are two different flavors of this local-search:*

- *Fleet-opt A: given two node-disjoint routes p and q associated to the respective vehicles $v_p$ and $v_q$. We exchange the vehicles, such that $v_q$ is associated to p and $v_p$ is associated to q.*
- *Fleet-opt B: we can replace a given vehicle $v_p$ associated to the route p by some different available vehicle $v_d$.*

**Definition 3 (Exchange).** *Consider two node-disjoint routes p and q that serve two distinct customers $i \in p$ and $j \in q$. We literally exchange the customers as follows. The edges $(i-1,i),(i,i+1) \in p$ are replaced by $(i-1,j),(j,i+1)$, and the edges $(j-1,j),(j,j+1) \in q$ are replaced similarly, by $(j-1,i),(i,j+1)$. Figure 3 illustrates this local search.*

**Definition 4 (Relocate).** *Given two node-disjoint routes p and q, and some customer i that belongs to p. We relocate the customer i to the route q, as follows. First, replace the edges $(i-1,i)$ and $(i,i+1)$ by $(i-1,i+1)$, and then replace the edge $(j,j+1) \in q$ by the edges $(j,i)$ and $i,j+1$. An illustration is presented in Fig. 4.*

**Definition 5 (2-opt).** *Pick two non-adjacent edges $(i,i+1)$ and $(j,j+1)$ from a fixed tour of a feasible solution, such that $i < j$. Replace both links by $(i,j)$ and $(i+1,j+1)$. Figure 5 illustrates this local search in a fixed tour.*

**Definition 6 (3-opt).** *Pick three non-adjacent edges $(i,j)$, $(k,l)$ and $(m,n)$. We can either delete two, or the three edges. In the former, we replace as in 2-opt. In the latter, consider the four non-isomorphic reconstructions of the tour illustrated in Fig. 6.*

The reader can appreciate that 3-opt is dominant, with cubic time-complexity in terms of the number of links.

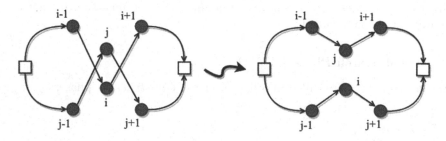

**Fig. 3.** Exchange

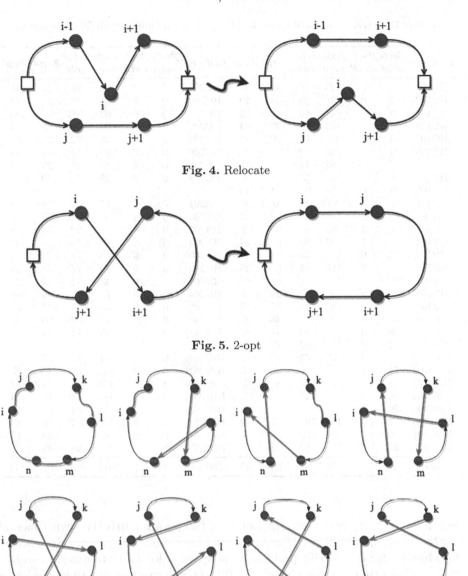

**Fig. 4.** Relocate

**Fig. 5.** 2-opt

**Fig. 6.** 3-opt

## 5    Numerical Results

In order to understand the effectiveness of our proposal, an extensive computational study was carried out using our *Main* algorithm versus the exact CPLEX solver, with an halting time of 10 h. Therefore, CPLEX returns either the globally

**Table 1.** Activity of the different local searches (instances with 50 customers).

| Instance | fleet-optA | fleet-optB | exch-ange | relo-cate | 2-opt | 3-opt | Instance | fleet-optA | fleet-optB | exch-ange | relo-cate | 2-opt | 3-opt |
|---|---|---|---|---|---|---|---|---|---|---|---|---|---|
| HC101 | 0 | 0 | 123 | 96 | 0 | 0 | HC201 | 0 | 0 | 28 | 26 | 12 | 0 |
| HC102 | 0 | 0 | 128 | 93 | 42 | 20 | HC202 | 0 | 0 | 45 | 46 | 28 | 24 |
| HC103 | 0 | 0 | 90 | 66 | 41 | 25 | HC203 | 0 | 0 | 92 | 102 | 50 | 50 |
| HC104 | 0 | 0 | 162 | 118 | 75 | 34 | HC204 | 0 | 0 | 64 | 57 | 40 | 28 |
| HC105 | 0 | 0 | 98 | 74 | 26 | 0 | HC205 | 0 | 0 | 36 | 40 | 29 | 18 |
| HC106 | 0 | 0 | 53 | 42 | 18 | 0 | HC206 | 0 | 0 | 45 | 51 | 36 | 24 |
| HC107 | 0 | 0 | 107 | 77 | 43 | 0 | HC207 | 0 | 0 | 40 | 38 | 25 | 19 |
| HC108 | 0 | 0 | 67 | 54 | 28 | 4 | HC208 | 0 | 0 | 39 | 40 | 30 | 20 |
| HC109 | 0 | 0 | 138 | 99 | 64 | 19 | | | | | | | |
| HR101 | 66 | 5 | 69 | 45 | 0 | 0 | HR201 | 0 | 0 | 31 | 40 | 29 | 14 |
| HR102 | 114 | 6 | 144 | 107 | 50 | 27 | HR202 | 0 | 0 | 97 | 140 | 74 | 53 |
| HR103 | 40 | 1 | 58 | 40 | 23 | 4 | HR203 | 0 | 0 | 71 | 92 | 58 | 41 |
| HR104 | 62 | 0 | 134 | 90 | 66 | 14 | HR204 | 0 | 0 | 48 | 55 | 32 | 22 |
| HR105 | 33 | 0 | 50 | 38 | 18 | 0 | HR205 | 0 | 0 | 59 | 78 | 52 | 33 |
| HR106 | 74 | 2 | 120 | 87 | 55 | 23 | HR206 | 0 | 0 | 49 | 76 | 38 | 26 |
| HR107 | 69 | 2 | 140 | 109 | 64 | 19 | HR207 | 0 | 0 | 42 | 51 | 29 | 26 |
| HR108 | 12 | 0 | 58 | 33 | 26 | 8 | HR208 | 0 | 0 | 71 | 85 | 46 | 36 |
| HR109 | 71 | 0 | 145 | 99 | 67 | 13 | HR209 | 0 | 0 | 48 | 61 | 45 | 32 |
| HR110 | 31 | 0 | 68 | 51 | 35 | 0 | HR210 | 0 | 0 | 122 | 155 | 90 | 63 |
| HR111 | 58 | 0 | 126 | 92 | 61 | 20 | HR211 | 0 | 0 | 55 | 71 | 44 | 33 |
| HR112 | 32 | 0 | 102 | 67 | 43 | 2 | | | | | | | |
| HRC101 | 43 | 4 | 87 | 43 | 41 | 4 | HRC201 | 0 | 0 | 53 | 37 | 27 | 6 |
| HRC102 | 38 | 1 | 80 | 46 | 37 | 7 | HRC202 | 0 | 0 | 86 | 57 | 40 | 9 |
| HRC103 | 18 | 1 | 65 | 37 | 27 | 5 | HRC203 | 2 | 1 | 158 | 122 | 74 | 27 |
| HRC104 | 26 | 0 | 78 | 47 | 34 | 2 | HRC204 | 0 | 0 | 185 | 132 | 76 | 19 |
| HRC105 | 29 | 1 | 87 | 45 | 39 | 3 | HRC205 | 0 | 0 | 7 | 58 | 36 | 12 |
| HRC106 | 47 | 0 | 152 | 91 | 71 | 12 | HRC206 | 0 | 0 | 170 | 122 | 86 | 17 |
| HRC107 | 19 | 0 | 63 | 38 | 28 | 5 | HRC207 | 0 | 0 | 62 | 44 | 29 | 15 |
| HRC108 | 36 | 0 | 110 | 69 | 52 | 8 | HRC208 | 0 | 0 | 98 | 69 | 49 | 9 |
| **Total** | **256** | **7** | **722** | **416** | **329** | **46** | **Total** | **2** | **1** | **819** | **641** | **417** | **114** |

optimum solution, or the best solution found so far after 10 h. The experimental analysis was carried out in a Home-PC (Intel Core i5 2.7 GHz). Since there are no benchmarks for our specific problem we adapted Solomon instances [15], adding penalties and time-windows, using $\omega = 0.3$. This means that the time-window is enlarged a factor 1.3, but a penalty is assumed in the last portion of the window. Since the HFVRPTW with penalties in delays is novel, we cannot perform a fair comparison with previous proposals. Instead, we study the effectiveness of our methodology with respect to the exact CPLEX solver, and the activity of the different local searches. Table 1 shows the activity of the five local searches of our VND. Exchange and Relocate have the largest activity, followed by 2-opt and 3-opt. Fleet-opt has the least activity. However, Fleet-opt A has considerable activity as well in many instances under study. Further experiments show that Fleet-opt B has large activity when the number of customers is increased to 100 and 200.

**Table 2.** CPLEX VS our GRASP/VND proposal (instances with 50 customers).

| CPLEX GRASP/VND | | | Difference | |
|---|---|---|---|---|
| Instance | Cost | Cost | Gap | Relative error |
| HC101 | 828.912 | 936.61 | 107.70 | 12.99% |
| HC102 | 871.274 | 887.47 | 16.20 | 1.86% |
| **HC103** | 994.16 | 887.86 | **−106.30** | **−10.69%** |
| **HC104** | 901.258 | 886.28 | **−14.98** | **−1.66%** |
| HC105 | 832.252 | 936.34 | 104.09 | 12.51% |
| HC106 | 805.756 | 937.79 | 132.03 | 16.39% |
| HC107 | 872.714 | 921.04 | 48.33 | 5.54% |
| HC108 | 903.65 | 936.91 | 33.26 | 3.68% |
| **HC109** | 1036.794 | 883.53 | **−153.26** | **−14.78%** |
| HC201 | 691.32 | 738.45 | 47.13 | 6.82% |
| HC202 | 645.58 | 750.25 | 104.67 | 16.21% |
| **HC203** | 828.57 | 706.99 | **−121.58** | **−14.67%** |
| **HC204** | 724.75 | 674.55 | **−50.20** | **−6.93%** |
| HC205 | 690.93 | 737.47 | 46.54 | 6.74% |
| **HC206** | 825.22 | 696.31 | **−128.91** | **−15.62%** |
| **HC207** | 843.96 | 716.15 | **−127.81** | **−15.14%** |
| **HC208** | 772.99 | 719.22 | **−53.77** | **−6.96%** |
| HR101 | 2475.672 | 2,577.31 | 101.63 | 4.11% |
| **HR102** | 2678.248 | 2,488.57 | **−189.67** | **−7.08%** |
| **HR103** | 2674.69 | 2,450.45 | **−224.24** | **−8.38%** |
| **HR104** | 2463.58 | 2,285.83 | **−177.75** | **−7.21%** |
| **HR105** | 2629.696 | 2,517.14 | **−112.56** | **−4.28%** |
| **HR106** | 2781.796 | 2,431.01 | **−350.78** | **−12.61%** |
| **HR107** | 2578.342 | 2,338.32 | **−240.02** | **−9.31%** |
| **HR108** | 2503.142 | 2,321.05 | **−182.09** | **−7.27%** |
| **HR109** | 2484.816 | 2,401.86 | **−82.95** | **−3.34%** |
| **HR110** | 2658.084 | 2,359.41 | **−298.68** | **−11.24%** |
| **HR111** | 2496.894 | 2,341.67 | **−155.22** | **−6.22%** |
| **HR112** | 2406.202 | 2,334.26 | **−71.94** | **−2.99%** |
| Average | − | − | **−155.37** | **−6.12%** |

Tables 2 shows the performance of our proposal with respect to CPLEX for 50 customers. The bold instances present negative gaps; this means that our GRASP/VND proposal outperforms CPLEX and, naturally, CPLEX could not find the globally optimum solution during 10 h. It is worth to remark that the average gap is negative, meaning that our proposal outperforms the solver.

Furthermore, the CPU time of our GRASP/VND solution ranges between seconds and five minutes in the worst cases.

## 6    Conclusions and Trends for Future Work

Operational researchers are engaged with modeling variations of the celebrated Vehicle Routing Problem (VRP), given its paramount importance and diverse applications. Here we introduced a novel Heterogeneous Fleet VRP with Time Windows (HFVRPTW) version, with penalties due to overtime. The HFVRPTW belongs to the class of $\mathcal{NP}$-Hard problems, since it subsumes the Traveling Salesman Problem. This result promotes the development of approximative algorithms. A GRASP/VND methodology is here proposed, using five different local searches. Numerical results suggest that the most simple local searches have more activity; further experiments illustrate that fleet-opt local search works when the number of customers is increased. The exact solution show limited applicability, where the optimality is reached only under small-sized instances.

As future work, we want to introduce key concepts of the VRP and variations in real-life metropolitan transportation systems. Further, we would like to explore novel local searches and study different versions of VNS ruled by probabilistic flow diagrams considering Markov chains.

**Acknowledgements.** This work is partially supported by Project ANII FCE_1_2019_1_156693 *Teoría y Construcción de Redes de Máxima Confiabilidad*, MATHAMSUD 19-MATH-03 *Rare events analysis in multi-component systems with dependent components* and STIC-AMSUD ACCON *Algorithms for the capacity crunch problem in optical networks*.

## References

1. Baldacci, R., Bartolini, E., Mingozzi, A., Roberti, R.: An exact solution framework for a broad class of vehicle routing problems. Comput. Manage. Sci. **7**(3), 229–268 (2010)
2. Bernal, J., Escobar, J.W., Linfati, R.: A granular tabu search algorithm for a real case study of a vehicle routing problem with a heterogeneous fleet and time windows. J. Ind. Eng. Manage. **10**(4), 646 (2017)
3. Cömert, S., Yazgan, H.R., Sertvuran, I., Şengüi, H.: A new approach for solution of vehicle routing problem with hard time window: an application in a supermarket chain. Sādhanā **42**(12), 2067–2080 (2017)
4. Dantzig, G.B., Ramser, J.H.: The truck dispatching problem. Manage. Sci. **6**(1), 80 (1959)
5. Garey, M.R., Johnson, D.S.: Computers and Intractability: A Guide to the Theory of NP-Completeness. W. H. Freeman & Co., New York (1979)
6. Hansen, P.: Variable neighborhood search. In: Handbook of Heuristics. Springer International Publishing (2018)
7. Jiang, J., Ng, K.M., Poh, K.L., Teo, K.M.: Vehicle routing problem with a heterogeneous fleet and time windows. Exp. Syst. Appl. **41**(8), 3748–3760 (2014)

8.  Karp, R.M.: Reducibility among Combinatorial Problems, pp. 85–103. Springer, Boston (1972). https://doi.org/10.1007/978-1-4684-2001-2_9
9.  Mor, A., Speranza, M.G.: Duality in nonlinear programming: a simplified applications-oriented development. 4OR - Q. J. Oper. Res. 18(2), 129–149 (2020)
10. Nagata, Y., Bräysy, O., Dullaert, W.: A penalty-based edge assembly memetic algorithm for the vehicle routing problem with time windows. Comput. Oper. Res. 37(4), 724–737 (2010)
11. Pessoa, A., Sadykov, R., Uchoa, E.: Enhanced branch-cut-and-price algorithm for heterogeneous fleet vehicle routing problems. Eur. J. Oper. Res. 270, 530–543 (2018)
12. Pop, P.C., Fuksz, L., Marc, A.H.: A variable neighborhood search approach for solving the generalized vehicle routing problem. In: Polycarpou, M., de Carvalho, A.C.P.L.F., Pan, J.-S., Woźniak, M., Quintian, H., Corchado, E. (eds.) HAIS 2014. LNCS (LNAI), vol. 8480, pp. 13–24. Springer, Cham (2014). https://doi.org/10.1007/978-3-319-07617-1_2
13. Resende, M., Ribeiro, C.: Optimization by GRASP. Springer (2016). https://doi.org/10.1007/978-1-4939-6530-4
14. Schneider, M., Sand, B., Stenger, A.: A note on the time travel approach for handling time windows in vehicle routing problems. Comput. Oper. Res. 40(10), 2564–2568 (2013)
15. Solomon, M.M.: On the worst-case performance of some heuristics for the vehicle routing and scheduling problem with time window constraints. Networks 16(2), 161–174 (1986)
16. Taha, A., Hachimi, M., Moudden, A. : A discrete bat algorithm for the vehicle routing problem with time windows. In: 2017 International Colloquium on Logistics and Supply Chain Management (LOGISTIQUA), pp. 65–70, April 2017
17. Taş, D., Jabali, O., Van Woensel, T.: A vehicle routing problem with flexible time windows. Comput. Oper. Res. 52, 39–54 (2014)

# Using K-Means and Variable Neighborhood Search for Automatic Summarization of Scientific Articles

I. Akhmetov[1,2(✉)] [iD], N. Mladenovic[3] [iD], and R. Mussabayev[1] [iD]

[1] Institute of Information and Computational Technologies, Pushkin str. 125, Almaty, Kazakhstan
i.akhmetov@ipic.kz
[2] Kazakh-British Technical University, Almaty, Kazakhstan
[3] Khalifa University of Science Technology: Abu Dhabi, Abu Dhabi, UAE
nenadmladenovic12@gmail.com
http://iict.kz, http://kbtu.kz

**Abstract.** This work presents a method for summarizing scientific articles from the arXive dataset using Variable Neighborhood Search (VNS) heuristics to automatically find the best summaries in terms of ROUGE-1 score we could assemble from scientific article text sentences. Then vectorizing the sentences using BERT pre-trained language model and augmenting the vectors with topic embeddings obtained by applying the K-means algorithm. Finally, training the Random Forest classification model to find sentences suitable for the summary and compile a summary from the selected sentences. The described algorithm produced summaries with high ROUGE-1 scores (0.45 on average), so we are heading for further developments on a larger dataset.

**Keywords:** Variable Neighborhood Search · VNS · K-means clustering · Text summarization · NLP · Random Forest Classifier

## 1 Introduction

As Isaac Newton once said regarding his significant contribution to science that "If I have seen further it is by standing on the shoulders of Giants" [1], meaning that his discoveries were made possible by the numerous works of scientists from as far as ancient times to his contemporaries. Nowadays, the Giants or researchers with their scientific articles contribution has grown so large and continues to grow at the accelerating rate with the information technologies developments. Standing on their shoulders became both dangerous and brought little use because we need tools to efficiently process a tremendous amount of information. Moreover, here come the methods of text summarization[1] with the

---

[1] Automatic text summarization - is a process of extracting the most important information from a text.

© Springer Nature Switzerland AG 2021
N. Mladenovic et al. (Eds.): ICVNS 2021, LNCS 12559, pp. 166–175, 2021.
https://doi.org/10.1007/978-3-030-69625-2_13

research in the area starting in 1958 [7] and bringing new numerous papers and methods every year since 2003 [2] when the large data sets for the purpose and necessary computing equipment became available.

We present an extractive summarization approach[2] using BERT [3] pre-trained language model for sentence embeddings and Random Forest algorithm [4] for classifier training.

As a superpower of our summarizer comes the Variable Neighborhood Search (VNS) [5,6], the latter technique will help us search for the best extractive summary available, allowing us to perform automatic sentence labeling avoiding manual workload.

We show how blending the modern contextual embedding method (BERT), Random Forest classification algorithm, and smart search heuristics (VNS) for data labeling can build a text summarization method that can, in perspective, achieve a new level of performance.

In Sect. 2, we give a short intro for the summarization task in Natural Language Processing (NLP). Section 3 describes the data we have used for our experiments and the methodology used. In Sect. 4, we present the results obtained, and Sect. 5 comes with the conclusion and sets out prospects for future work.

## 2 Related Work

Many approaches have been developed since the first paper on the text summarization subject was published by Luhn in 1958 [7], developing from the purely statistical to more recent machine learning (ML) [8] and contemporary Deep Learning methods [9–11].

Generally, we can classify the text summarization methods as follows:

1. **Input**
   (a) Single document – summarization of one single document as a whole.
   (b) Multi-document – using a series of documents related to a common subject, but occurring simultaneously. It can be used in the literature review process of scientific work to receive short and concise information on a subject reducing redundancy.
2. **Output**
   (a) Extractive – summary uses the exact sentences from the source text without paraphrasing or combining them. This type of summary resembles bullet points to anchor for the main in-formation and often lacks transitive phrases and sentences for smoothing the text.
   (b) Abstractive – summarizes by the own words of a person who read the article. The method is more complex than Extractive and involves sentence templates or advanced Natural Language Generation (NLG) models.

---

[2] The source code is available on GitHub at https://github.com/iskander-akhmetov/ Using-k-means-and-Variable-Neighborhood-Search-for-automatic-summarization-of-scientific-articles/.

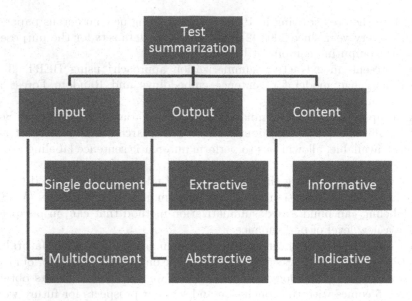

**Fig. 1.** Text summarization methods classification [12,13].

3. **Content**
   (a) Informative – the summary captures all the main information from the source text such that there is no need to read it after reading the summary.
   (b) Indicative – the summary is a kind of teaser motivating a person to read the whole article if it seems relevant to the current information query (Fig. 1).

The silver bullet or a superpower we will use in this paper is the Variable Neighborhood Search algorithm first introduced by N. Mladenovic [14] as a local search heuristic for solving of the minimum sum of squares problem in the clustering algorithms. Put merely, VNS takes the initial solution of a problem and iteratively changes the volume of change when no improvement to finding the objective function optima occurs and fixing the best result.

The text sentences will be vectorized by the pre-trained Bidirectional Encoder Representations from Transformers (BERT) English language model. BERT is designed to pre-train deep bidirectional unlabeled text representations. The pre-trained BERT model can be fine-tuned with just one additional output layer to create state-of-the-art models for a wide range of NLP tasks, such as summarization and question answering. BERT conceptual simplicity and empirical power fullness allow obtaining new state-of-the-art results [3].

## 3   Methods and Data

### 3.1   Data

Firstly introduced along with the PubMed dataset in 2018 [15] it 215 K arXiv.org repository scientific articles in English from the domains of physics, math, and other quant fields. The dataset articles contain abstracts, main articles, a list of sections, and main article texts divided by sections. The average length of articles and abstracts is 4938 and 220 words, respectively.

## 3.2   Methods

**Attaining Best Achievable ROUGE-1 Score Using VNS.** Using the original article abstracts in our dataset as a reference, try to assemble, with the VNS technique's help, the best summary in terms of ROUGE-1 metric from article sentences.

Applying VNS here is logically derived from the fact that finding the best possible summary out of the text sentences by running through all possible combinations is impractical due to the hyper exponential complexity of this approach:

$$\binom{N_t}{N_a} = \frac{N_t!}{N_a!(N_t - N_a)!} \tag{1}$$

where $N_a$ and $N_t$ - are the number of sentences in summary and text, respectively. While VNS gives a rather simpler alternative, which can provide us a satisfactory solution for a reasonable amount of time.

VNS is a framework for building heuristics (meta-heuristic), which exploits the idea of systematical initial solution neighborhood change to find optimums of the objective function to select from [23].

VNS exploits systematically the following observation facts [23]:

1. A local minimum concerning one neighborhood structure is not necessarily so for another.
2. A global minimum is a local minimum for all possible neighborhood structures.
3. For many problems, local minima to one or several neighborhoods are relatively close to each other.

---

*Initialization.* Select the set of neighborhood structures $\mathcal{N}_k$, for $k = 1, \ldots, k_{max}$, that will be used in the search; find an initial solution $x$; choose a stopping condition;
*Repeat* the following sequence until the stopping condition is met:
(1) Set $k \leftarrow 1$;
(2) *Repeat* the following steps until $k = k_{max}$:
    *(a) Shaking.* Generate a point $x'$ at random from the $k$th neighborhood of $x$ ($x' \in \mathcal{N}_k(x)$);
    *(b) Move or not.* If this point is better than the incumbent, move there ($x \leftarrow x'$), and continue the search with $\mathcal{N}_1$ ($k \leftarrow 1$); otherwise, set $k \leftarrow k+1$;

---

**Fig. 2.** General VNS algorithm pseudo code.

In our case (Fig. 2), using VNS terminology:

- Initial solution - is the first, usually random approximation of the objective function. We initialize our search for solution by a random set of sentences $x$ in $\mathcal{N}_k = \binom{N_t}{N_a}$ space of possible neighborhood structures.
- Shaking - is the process of systematical modification of the initial solution to the extent specified by the $k_{max}$ parameter.
- Incumbent solution - is the best current solution achieved after shaking.
- Stop condition - the cycle is limited by 5000 iterations or 60 seconds time. If no ROUGE-1 score improvement occurs after 700 consecutive iterations, the cycle breaks.

We have randomly selected 100 articles from the arXive dataset and for each of them performed the following steps in a cycle:

1. **Initial solution** - Calculating the ROUGE-1 [16] score for an initially random set of sentences from the original document,
2. **Shaking** - Make changes starting from replacing one randomly selected sentence by a new one from the text up to $k_{max}$ sentence replacement if no Rouge-1 score improvement occurs. The maximum amount of the changes is $k_{max}$ parameter.
3. **Incumbent solution** - Recalculate ROUGE-1 score and fix the result if it is better than the initial score and reset the $k$ to one sentence, or if no improvement happens, gradually increase the $k$ up to $k_{max}$.

**Summarization Method.** Our summarization method is supervised and will consist of the following steps:

1. Creating the dataset to classify document sentences whether they belong to the best available summary sentence set or not. The algorithm is described in the "Attaining best achievable ROUGE score using VNS" section. Here we add labels to each sentence for our classification algorithm.
2. BERT vectorization: using the pre-trained BERT-Base, Cased [3] along with the PyTorch Pre-trained BERT [22] package for Python. We vectorize sentences to obtain 768 elements feature vector for every sentence, specifically the BERT model's 12th layer, as it carries the most variance. As a result of this step, we 23 K sentence BERT embeddings from a sample of 100 arXive dataset articles labeled as to whether the sentence is good for the summary or not (Figs. 3 and 4).
3. K-means clustering with VNS: applying basic K-means clustering algorithm to obtain a quasi-topic attribution for the sentences and thus converting the data to a new vector space, using VNS:
   (a) Clustering the sentence vectors for 100 clusters (the number we chose arbitrarily) using traditional K-means algorithm
   (b) Enhance the quality metric (sum of squared distances from elements to centroids) for clustering using the VNS algorithm

(c) Calculate the similarity of each sentence vector to each cluster center

$$A = \left(a_{i,j}\right)$$
$$B = \left(c_{j,k}\right) \tag{2}$$
$$C = A \times B$$

where A - BERT vectors of the sentences in the dataset, B - centroid vectors, C – distance matrix from each matrix to each centroid, i - number of sentences, j - length of BERT vector used, 768 in our case, k - number of clusters, 100 in our case.

(d) For each sentence vector, calculate the probability of attribution to each of 100 clusters by dividing the corresponding cosine similarities by the sum of similarities.

$$Topic_{prob} = \frac{C}{C_{sum}}$$
$$C_{sum} = \left( \sum_{n=1}^{i} C_{n,1}, \sum_{n=1}^{i} C_{n,2}, ..., \sum_{n=1}^{i} C_{n,k} \right) \tag{3}$$

where C - are the vectors obtained on the previous step.

In this step, we obtain a topic embedding for each sentence.

4. Sentence classification: using the Random Forest Classifier algorithm, we classify sentences inputting topic embeddings as X and labels indicating whether the sentence is good for summary(1) or not(0) which is by nature a supervised binary classification task.

5. Summary assembly: applying classifier model trained on previous step to a text sentences which are vectorized by BERT model, we are able to select the same number of sentences from the text as in original article abstract, and which are best for the summary, and assemble a summary from them.

6. Evaluation of the algorithm was made with ROUGE-1 metric of the produced summary against the golden summary.

## 4    Results

Applying the VNS technique described above for labeling article sentences whether they are good for summary or not on average we have got a ROUGE-1 metric of 0.55 for arXive y, whereas the best contemporary methods employing sophisticated neural network architectures [9–11] are able to achieve ROUGE-1 of just 0.45 on arXive dataset. So there is definitely a room for improvement for the summarization methods and techniques.

The Random Forest Classifier training was performed on a train/test split of 0.33, achieving accuracy on the training sample of 0.99 and 0.95 on train and test. High results are not surprising as the data is highly imbalanced for the ratio of 1:20 in favor of the majority class, which are the sentences that are not suitable for the summary (Tables 1, 2 and 3).

**Table 1.** Comparison of the contemporary leader text summarization models.

| Class | Model | ROUGE-1 |
|---|---|---|
| Extractive | SumBasic [15–17] | 29.47 |
| | LexRank [15, 18] | 33.85 |
| | LSA [15, 19] | 29.91 |
| Abstractive | Attn-Seq2Seq [15, 20] | 29.30 |
| | PEGASUS [9] | 44.70 |
| | Pntr-Gen-Seq2Seq [15, 21] | 32.06 |
| | Discourse-att [15] | 35.8 |

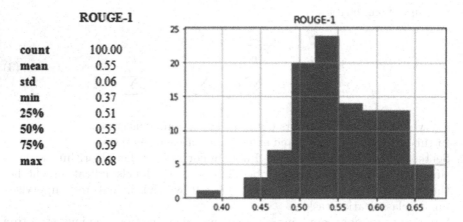

**Fig. 3.** The best achievable ROUGE-1 score on a 100 sample of arXive dataset articles.

**Table 2.** Confusion matrix of the classification results on test sample.

| | | Predicted | |
|---|---|---|---|
| | | 0 | 1 |
| Actual | 0 | 7,264 | 1 |
| | 1 | 367 | 36 |

The confusion matrix shows that the classifier is reasonably good at detecting the non-summary sentences miss classifying for negative, just a single sentence. It is not that perfect at detecting the summary sentences with a false-negative rate of 0.91.

**Table 3.** Classification report.

| | |
|---|---|
| Precision | 0.97 |
| Recall | 0.09 |
| Balanced accuracy | 0.55 |
| F1 | 0.16 |

The classification report shows the classifier's weak side – the false-negative rate and, thus, a low F1 score. We used the balanced accuracy metric here as the dataset is highly imbalanced, and the standard accuracy metric is misleading, as mentioned above.

Despite decent classification results, the algorithm was able to produce summaries with a high ROUGE-1 score. Evaluating the proposed summarization method on the combined train and test samples, we observe that 75% of the generated summaries achieved a ROUGE-1 score of more than 0.41, and 50% of summaries got a result greater than 0.49. On average, the method yields in 0.45, and so it seems to be a good beginning for further algorithm development and training it on a significant amount of texts.

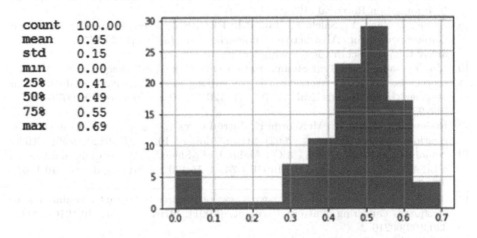

**Fig. 4.** Evaluation of the summarizer on combined test/train set.

## 5 Conclusion

In this work, we have tried to show a promising approach to using the VNS for NLP's summarization task. We have applied the technique for the finding of the best possible ROUGE-1 score on the extractive summaries. Considering the drawbacks of the resulting algorithm in its classifier part, we aim to solve the problem and try our algorithm on a much bigger data set.

**Acknowledgement.** This work was supported by the Science Committee of RK, under the grants AP08856034, AP09058174, BR05236839.

# References

1. Knowles, E.: Oxford Dictionary of Quotations. Oxford University Press, Oxford (2001)
2. Graff, D., Cieri, C.: English Gigaword - Linguistic Data Consortium. Linguistic Data Consortium (2003)
3. Devlin, J., Chang, M.-W., Lee, K., Toutanova, K.: BERT: Pre-training of Deep Bidirectional Transformers for Language Understanding (2018)
4. Breiman, L.: Random forests. Mach. Learn. (2001). https://doi.org/10.1023/A: 1010933404324
5. Hansen, P., Mladenović, N.: Variable neighborhood search. In: Handbook of Heuristics (2018). https://doi.org/10.1007/978-3-319-07124-4_19
6. Hansen, P., Mladenović, N., Moreno Pérez, J.A.: Variable neighbourhood search: methods and applications. Ann. Oper. Res. (2010). https://doi.org/10.1007/ s10479-009-0657-6
7. Luhn, H.P.: The automatic creation of literature abstracts. IBM J. Res. Dev. **2**, 159–165 (1958). https://doi.org/10.1147/rd.22.0159
8. Kupiec, J., Pedersen, J.: A trainable document summarizer. In: Proceedings of the 18th Annual International ACM SIGIR Conference on Research and Development In Information Retrieval (1995)
9. Zhang, J., Zhao, Y., Saleh, M., Liu, P.J.: PEGASUS: Pre-training with Extracted Gap-sentences for Abstractive Summarization. (arXiv:1912.08777v1 [cs.CL]). arXiv Comput. Sci. https://doi.org/arXiv:1912.08777v1
10. Liu, Y., Lapata, M.: Text Summarization with Pretrained Encoders (2019)
11. Lloret, E., Plaza, L., Aker, A.: The challenging task of summary evaluation: an overview. Lang. Resour. Eval. **52**, 101–148 (2018). https://doi.org/10.1007/s10579-017-9399-2
12. Radev, D.R., Hovy, E., McKeown, K.: Introduction to the special issue on summarization. Comput. Linguist. (2002). https://doi.org/10.1162/089120102762671927
13. Abualigah, L., Bashabsheh, M.Q., Alabool, H., Shehab, M.: Text summarization: a brief review. Stud. Comput. Intell. **874**, 1–15 (2020). https://doi.org/10.1007/ 978-3-030-34614-0_1
14. Hansen, P., Mladenović, N.: J-Means: a new local search heuristic for minimum sum of squares clustering. Pattern Recognit. (2001). https://doi.org/10.1016/S0031-3203(99)00216-2
15. Cohan, A., et al.: A discourse-aware attention model for abstractive summarization of long documents. In: Proceedings of the 2018 Conference of the North American Chapter of the Association for Computational Linguistics: Human Language Technologies, Volume 1 (Long Papers), vol. 2, pp. 615–621 (2018). https://doi.org/10. 18653/v1/n18-2097
16. Lin, C.-Y.: ROUGE: a package for automatic evaluation of summaries. In: Text Summarization, pp. 74–81 (2004)
17. Vanderwende, L., Suzuki, H., Brockett, C., Nenkova, A.: Beyond SumBasic: task-focused summarization with sentence simplification and lexical expansion. Inf. Process. Manage. (2007). https://doi.org/10.1016/j.ipm.2007.01.023
18. Erkan, G., Radev, D.R.: LexRank: graph-based lexical centrality as salience in text summarization. J. Artif. Intell. Res. (2004). https://doi.org/10.1613/jair.1523

19. Jezek, K., Steinberger, J., Ježek, K.: Using latent semantic analysis in text summarization and summary evaluation. In: Proceedings of the 7th International Conference ISIM (2004)
20. Nallapati, R., Zhou, B., dos Santos, C., Gulçehre, Ç., Xiang, B.: Abstractive text summarization using sequence-to-sequence RNNs and beyond. In: CoNLL 2016–20th SIGNLL Conference on Computational Natural Language Learning, Proceedings (2016). https://doi.org/10.18653/v1/k16-1028
21. See, A., Liu, P.J., Manning, C.D.: Get to the point: summarization with pointer-generator networks. In: Proceedings of the 55th Annual Meeting of the Association for Computational Linguistics (Volume 1: Long Papers), pp. 1073–1083 (2017). https://doi.org/10.18653/v1/P17-1099
22. Wolf, T., et al.: HuggingFace's Transformers: State-of-the-art Natural Language Processing (2019)
23. Burke, E., Kendall, G.: Search Methodologies: Introductory Tutorials in Optimization and Decision Support Techniques. Springer, New York (2014). https://doi.org/10.1007/978-1-4614-6940-7

# BVNS Approach for the Order Processing in Parallel Picking Workstations

Abdessamad Ouzidan[1], Eduardo G. Pardo[2(⊠)], Marc Sevaux[1],
Alexandru-Liviu Olteanu[1], and Abraham Duarte[2]

[1] Lab-STICC, UMR 6285 CNRS, Université de Bretagne-Sud, Lorient, France
{abdessamad.ouzidan,marc.sevaux,alexandru.olteanu}@univ-ubs.fr
[2] Universidad Rey Juan Carlos, Madrid, Spain
{eduardo.pardo,abraham.duarte}@urjc.es

**Abstract.** The Order Processing in Parallel Picking Workstations is an optimization problem that can be found in the industry and is related to the order picking process in a warehouse. The objective of this problem is to minimize the number of movements of goods within a warehouse in order to fulfill the demand made by the customers. The goods are originally stored in containers in a storage location and need to be moved to a processing area. The processing area is composed of several identical workstations. We are particularly interested in minimizing the time needed to fulfill all demands, which corresponds to the highest number of container movements to any given workstation. This problem is $\mathcal{NP}$-Hard since it is a generalization of the well-known Order Processing in Picking Workstations which is also known to be $\mathcal{NP}$-Hard. In this paper, we provide a mathematical formulation for the problem and additionally, due to its hardness, we have also developed several heuristic procedures based on Variable Neighborhood Search, to tackle the problem. The proposed methods have been evaluated over different sets of instances.

**Keywords:** Order picking · Parallel workstations · Variable Neighborhood Search · Integer Linear Programming

## 1 Introduction

The Order Processing in Parallel Picking Workstations (O3PW) is an optimization problem that occurs in warehouses during the order picking task. This task is just one of the duties related to the supply chain and it involves the management of activities, information, and people within the warehouse [11,14]. Other related tasks are receiving, storing, and shipping goods to the customers. The importance of the order picking is well documented and some studies indicate that the picking process can be responsible of up to 55% of the operational costs in the warehouse [6,26].

Problems within the order picking category can be divided in two main subcategories: parts-to-picker or picker-to-parts [4,5]. If the containers which hold

© Springer Nature Switzerland AG 2021
N. Mladenovic et al. (Eds.): ICVNS 2021, LNCS 12559, pp. 176–190, 2021.
https://doi.org/10.1007/978-3-030-69625-2_14

the goods are moved to the workstations in the processing zone, avoiding any travel time for the picker, we are in the parts-to-picker category [21]. On the other hand, the picker-to-parts category includes problems where the picker travels through the warehouse to collect the items demanded by the customers [15,18,19]. In both previous situations, every order can be assigned to a picker and collected in isolation [9] or packed together with other orders (order batching) [4,5] and collected at the same time as the orders in the same batch [2,16,18]. Also, some warehouses run with only one operator while others have more than one [4,5,17,25] and, additionally, the warehouse can be manual or automated as it was introduced in [4] and later adapted in [22]. We refer the reader to [4,24] for surveys which review the previously described family of problems. In Fig. 1 we present the main categories aforementioned and reported in [22].

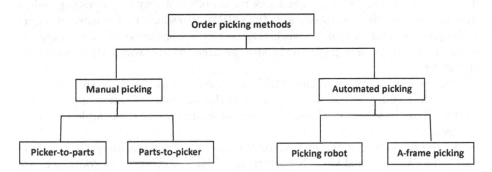

**Fig. 1.** Adaptation of the classification of order-picking methods previously proposed in [4].

In this paper we tackle the O3PW problem, which can be classified in either the manual or the automated picking category and, more specifically, within the parts-to-picker or picking robots subcategory, depicted in Fig. 1. The O3PW is a generalization of the Order Processing in Picking Workstations (OPPW), since the former considers many workstations while the latter considers only one. The OPPW was tackled in [10] where it was proved to be $\mathcal{NP}$-Hard. Also, the authors of this paper proposed a mathematical formulation and a heuristic procedure for solving it. Another variant of the OPPW and also closely related to the O3PW, named Mobile Robot based Order-Picking, uses containers which have multiple types of products, while the containers in the OPPW and O3PW have only one kind of product. The MROP variant was tackled in [3], where the authors also proved the problem to be $\mathcal{NP}$-Hard and proposed another heuristic procedure to solve it.

As far as we know, the O3PW, has never been tackled in the literature despite of its practical interest, since it illustrates real industry scenarios better than the OPPW. However, it is important to notice that, since the O3PW is a generalization of the OPPW, the O3PW is also $\mathcal{NP}$-Hard, since, reducing the number of picking workstations to one makes the O3PW identical to the OPPW, which is known to be $\mathcal{NP}$-Hard. Therefore we can conclude that the O3PW is at least as difficult to solve as OPPW.

The rest of the paper is structured in the following way. We begin by describing the O3PW problem in Sect. 2 followed by proposing a mathematical formulation in Sect. 3 for the O3PW. Section 4 contains several resolution approaches, such as a constructive heuristic and a Basic Variable Neighborhood Search [20], which are then tested and compared to each other and to the mathematical model in Sect. 5. Finally, Sect. 6 concludes this contribution with several remarks and perspectives for future work.

## 2   Problem Definition

The O3PW studied in this paper looks for an efficient way of processing orders arriving to a warehouse. Specifically, the O3PW looks for: i) a sequence of orders to be processed on each of the workstations of the warehouse and ii) a sequence of containers to be brought from the storage area to each workstation, to satisfy the orders.

The general overview of the O3PW is illustrated in Fig. 2. In this case, the warehouse is composed of: a storage area with containers of goods, a processing area with multiple workstations, and a shipping area where the completed orders are placed.

In the beginning of the picking process, all workstations are empty, so several orders are assigned to each workstation. Then, the necessary containers are brought from the storage area to each workstations to satisfy the orders. New orders can be inserted afterwards when one or multiple slots are free (i.e., once orders have been completed and sent to the shipping area). Notice that when a new order enters a workstation, it can be served by the container already present in this workstation. Once the container is in a workstation, one or more items can be extracted from the container to satisfy any of the orders currently at the workstation. The container is then returned to its original position in the storage area. No more items can be then extracted from this container unless it is retrieved again from the storage area and moved to the processing area (either to the same or to a new workstation). Notice that, each container can hold only one kind of item (single-item containers) and, additionally, containers can be retrieved either manually or automatically.

In this paper we assume that there are enough items to satisfy the orders being processed in the warehouse. As far as the processing area is concerned, it contains several workstations which are all identical. A workstation can hold only one container at a time and can manage simultaneously a number of orders which is determined by a parameter which is set depending on the problem instance. Notice, that this maximum capacity might be different from one instance to another. Also, an order does not have a maximum capacity limit for either products or items. We also suppose that there are no stock restrictions and that all the orders can be satisfied. Additionally, we suppose that containers cannot be transferred between workstations, and that an order which enters a workstation cannot be removed nor assigned to another workstation, until it is fully processed. Finally, let us remark that in this paper we are considering an offline version of the problem, where all the client orders are available at the beginning of the optimization process.

Following the previous remarks, the objective function of the O3PW is defined as minimizing the largest number of moves of containers of any workstation. In other words, the objective function can be also defined as the minimization of time needed to fulfill all orders when handling these order in parallel across multiple workstations.

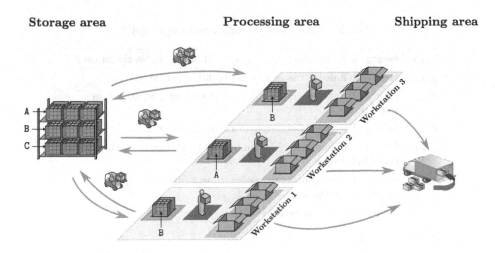

**Fig. 2.** O3PW general overview.

## 3    Mathematical Formulation

In order to properly define the O3PW, in this section, we provide a mathematical formulation for the problem. Let us consider the parameters and variables given in Table 1 and Table 2, respectively, for the proposed model.

The retrieval process can be organized in multiple steps. At every step, several moves of containers are performed in order to complete at least one order (notice that when performing those moves, more than one order can benefit from them, since they might share the same products. Therefore, more than one order might be processed at the same step). As $T$ is the maximum number of order slots available in a workstation, we can define an upper bound $K$ in the number of steps needed as $K = N - T + 1$, that guarantees that all orders will be processed. Given the previous definitions, we propose the following Integer Linear Programming formulation for the O3PW:

**Table 1.** Parameters of the mathematical model.

| | |
|---|---|
| $N$ | Number of orders to prepare ($i = 1, ..., N$) |
| $M$ | Number of different containers ($j = 1, ..., M$) |
| $L$ | Number of workstations in the processing area ($l = 1, ..., L$) |
| $T$ | Size of a workstation (number of available order slots) |
| $A$ | Customer demands represented as a matrix of size $N \times M$ where position $a_{i,j} = 1$ if item $j$ is demanded by order $i$ and $a_{i,j} = 0$ otherwise |
| $K$ | Number of processing steps ($k = 1, ..., K$) |

**Table 2.** Variables of the mathematical model.

| | |
|---|---|
| $x_{j,k,l}$ | *binary*: 1 if container $j$ is brought at step $k$ to workstation $l$ |
| $y_{i,k,l}$ | *binary*: 1 if order $i$ is in the workstation $l$ at step $k$ |
| $z_{i,j,k,l}$ | *binary*: 1 if order $i$ is served with container $j$ at step $k$ in workstation $l$ |
| $w_{i,k}$ | *binary*: 1 if order $i$ enters a workstation at step $k$ |
| $\delta_{j,k,l}$ | *binary*: 1 if container $j$ is the last to be retrieved at step $k$ for the Workstation $l$ |
| $\lambda_{k,l}$ | *binary*: 1 if no containers are retrieved from the storage area at step $k$ for the workstation $l$ |
| $s$ | *integer*: makespan of workstations |

$$\min s \tag{1}$$

s.t. :

$$\sum_{i=1}^{N} y_{i,k,l} \leq T \qquad k = 1, .., K; l = 1, ..., L \tag{2}$$

$$\delta_{j,1,l} \leq x_{j,1,l} \qquad j = 1, .., M; l = 1, ..., L \tag{3}$$

$$\delta_{j,k,l} \leq x_{j,k,l} + \lambda_{k,l} \qquad j = 1, .., M; k = 2, .., K; l = 1, ..., L \tag{4}$$

$$\sum_{j=1}^{M} \delta_{j,k,l} = 1 \qquad k = 1, .., K; l = 1, ..., L \tag{5}$$

$$\sum_{j=1}^{M} x_{j,k,l} \geq 1 - \lambda_{k,l} \qquad k = 2, .., K; l = 1, ..., L \tag{6}$$

$$\sum_{j=1}^{M} x_{j,k,l} \leq C(1 - \lambda_{k,l}) \qquad k = 2, .., K; l = 1, ..., L \tag{7}$$

$$\delta_{j,k,l} \geq \delta_{j,k-1,l} + \lambda_{k,l} - 1 \qquad j = 1, .., M; k = 2, .., K; l = 1, ..., L \tag{8}$$

$$z_{i,j,1,l} \leq x_{j,1,l} \qquad\qquad i = 1,..,N; j = 1,..,M; l = 1,...,L \qquad (9)$$

$$z_{i,j,k,l} \leq x_{j,k,l} + \delta_{j,k-1,l} \quad i = 1,..,N; j = 1,..,M; k = 2,..,K; l = 1,...,L \tag{10}$$

$$z_{i,j,k,l} \leq y_{i,k,l} \qquad\qquad i = 1,..,N; j = 1,..,M; k = 1,..,K; l = 1,...,L \tag{11}$$

$$\sum_{l=1}^{L}\sum_{k=1}^{K} z_{i,j,k,l} \geq a_{i,j} \qquad i = 1,..,N; j = 1,..,M \tag{12}$$

$$w_{i,1,l} \geq y_{i,1,l} \qquad\qquad i = 1,..,N; l = 1,...,L \tag{13}$$

$$w_{i,k,l} \geq y_{i,k,l} - y_{i,k-1,l} \quad i = 1,..,N; k = 2,..,K; l = 1,...,L \tag{14}$$

$$\sum_{l=1}^{L}\sum_{k=1}^{K} w_{i,k,l} = 1 \qquad i = 1,..,N \tag{15}$$

$$s \geq \sum_{j=1}^{M}\sum_{k=1}^{K} x_{j,k,l} \qquad l = 1,..,L \tag{16}$$

$$x_{j,k,l}, \delta_{j,k,l} \in \{0,1\} \qquad j = 1,..,M; k = 1,..,K; l = 1,...,L \tag{17}$$

$$y_{i,k,l}, w_{i,k,l} \in \{0,1\} \qquad i = 1,..,N; k = 1,..,K; l = 1,...,L \tag{18}$$

$$\lambda_k \in \{0,1\} \qquad\qquad k = 2,..,K \tag{19}$$

In this Integer Linear Programming formulation, the objective function is to minimize the maximum number of container moves made to a workstation. Constraints (2) guarantee that the number of orders being processed in a given workstation, at a particular step, does not exceed the number of available slots $T$. Constraints (3) to (8) indicate the last retrieved container at step $k$ for workstation $l$. Particularly, if some containers are retrieved from the storage area at step $k$ for workstation $l$ ($\lambda_k, l = 0$) then, the model will set one of these containers as the last one to be retrieved, otherwise when no containers are retrieved for workstation $l$ at step $k$ ($\lambda_k, l = 1$), the last container of this step will be identical to the last one brought at the previous step $k - 1$. In fact, the orders that will be assigned to the workstation $l$ at a step $k$, will be served from the last retrieved container at step $k - 1$ if its items are needed. The constant $C$ in the formulation must be a value big enough to satisfy the concerned constraints, for instance $C = N \times M$. Constraints (9) to (12) mean that every order should be completely processed according to its demand. Notice that, these constraints should only be taken into consideration when $a_{i,j}$ of matrix $A$ is equal to one. Constraints (13), (14), and (15) guarantee that, when an order enters a workstation, it is not removed from the slot until it is fully processed. Particularly, constraints (15) indicate that an order can only enter once across all workstations.

Further than mathematically formulating the O3PW problem, we expect that this model can be used within a solver to efficiently handle small-sized instances of the problem. Unfortunately, since we are facing a $\mathcal{NP}$-hard problem, other techniques, as the ones that we present next, might be more suitable for large instances.

# 4    Heuristic Approaches

The hardness of the O3PW along with the size of the real industrial instances for this problem, has led us to the use of heuristics and metaheuristics in order to provide high quality solutions for large-sized instances. Also, it is important to state that, when solving real problem instances, the quality of the solution and the time needed to find it are both equally important.

In this section, we propose the use of a constructive heuristic which will be used later as the procedure to produce a starting solution for the Basic Variable Neighborhood Search (BVNS) algorithm. These approaches will be described respectively in Sects. 4.1 and 4.2.

## 4.1    Constructive Heuristic

A constructive heuristic consists of a method of building a feasible solution for an optimization problem, that can be used as a final solution, or as a initial starting point for other heuristics or metaheuristics.

In order to design a constructive procedure for the problem, let us remind that a feasible solution for the O3PW consists of a sequence of orders to be processed on each workstation, and a sequence of containers to be brought from the storage area to each workstation. Also, an instance of the O3PW is composed of a set of orders to be processed and a predefined number of workstations $w_{max}$. In addition, all workstations are identical, which means they have the same number of slots $s_{max}$ to process orders, and to store containers (in this case, the space for containers is restricted to one). In this section, we propose a constructive procedure for the O3PW that will provide a solution used as the initial point for a BVNS method.

The constructive heuristic starts by assigning to every available workstation a randomly selected order. At this stage, every workstation has one order to process. Then, we randomly choose a workstation and we assign additional orders to it, as long as there are available slots, by choosing the most similar orders among the candidate orders awaiting to be processed. We evaluate this similarity by counting the number of products from the candidate order that appear in any order already assigned to the workstation. Once all slots have been filled, we pass onto another workstation and we repeat the process until all the workstations have no more free slots.

After completing the first step, all workstations have one or more orders assigned to their slots and the remaining orders are waiting to be assigned. Next, we need to start retrieving containers from the storage area in order to satisfy the demand made by the orders, and consequently to free slots for new orders. To determine the sequence of moves for a particular workstation, we select the order with the fewest remaining products and then we retrieve the products randomly, by moving containers from the storage area to the workstation one by one. Notice that, when a demanded product has been retrieved, other orders on the same workstation can also benefit from it. Once the considered order is completed and ready for shipping, it is sent to the shipping area and another order is assigned to the available slot. The first time we perform the order retrieving process we select one workstation at random, however during the following iterations, when one or multiple orders are completed, we select the workstation with the smallest number of moves already made. In this way, the workload given through the number of containers retrieved from the storage area is balanced across all workstations. The process is repeated until all orders have been processed. Note that any ties in this heuristic are broken at random.

## 4.2   Basic Variable Neighborhood Search

We propose the use of Variable Neighborhood Search (VNS) methodology in order to improve the solutions provided by the constructive procedure presented in Sect. 4.1. VNS is a metaheuristic procedure based on the idea of changes of neighborhood. It was proposed by Pierre Hansen and Nenad Mladenovic in 1997 [20] as a general methodology to solve hard optimization problems [13]. VNS has been widely extended with different algorithmic variants. The main extensions are the Reduced VNS, the Basic VNS, the Variable Neighborhood Descent and the General VNS. See [12] for a recent and thorough review. Other recent approaches are the Multi-Objective VNS [7], the Variable Formulation Search [23], and the Parallel VNS [8].

Among the different variants of VNS we have chosen the BVNS as the specific method to tackle the O3PW, since it finds a balance between stochastic and deterministic exploration of the space search. In Algorithm 1 we present the pseudocode of the BVNS algorithm. BVNS receives three input parameters: an initial solution $s$ (provided by our constructive procedure); the maximum number of neighborhoods to be explored ($k_{max}$); and $r_{max}$, which denotes the maximum number of non-improving iterations the algorithm is allowed to perform. Also, the BVNS procedure has three main steps: the Shake procedure (step 5), the LocalSearch procedure (step 6); and the neighborhood change procedure (represented here by steps 7 to 12). Next, we describe in detail each of them:

- The Shake procedure performs $k$ perturbations to the current solution $s$ in order to escape from the current basin of attraction. In this case, each perturbation consist of changing the assignment of and a randomly selected order from its current workstation to a different one. The resulting solution ($s'$) is provided as an input parameter to the LocalSearch procedure.
- The LocalSearch procedure tries to improve the solution by completely exploring a neighborhood until a local optimum is reached. The neighborhood is generated by performing a particular move to a solution. We consider here two different moves and, therefore, two different local search procedures. The first local search procedure ($LS_1$) is based the insertion move used by the Shake procedure. The second local search procedure ($LS_2$) is based on a swap move which consist of selecting two orders assigned to different workstations and interchanging their assignments. For both local search procedures, performing a move requires recalculating the sequence in which orders will enter the workstations and also the sequence of containers to satisfy the orders (as we described in the Sect. 4.1).

– The neighborhood change procedure compares the improved solution $(s'')$ provided by the LocalSearch with the best previous solution found $(s)$. If $(s'')$ is better than $(s)$, then $(s'')$ replaces $(s)$ and will be used as a starting point for the next iteration of the algorithm. Additionally, the value of the variable $k$ (which indicates the number of perturbations to be performed in the Shake procedure) will be set to 1. Otherwise, if there is no improvement, the method increases the value of $k$ by one.

The three main steps of BVNS are repeated until the value of $k$ equals $k_{max}$. The whole procedure is run again from the beginning, for at most $r_{max}$ times but starting each time from the new best solution found. Notice that whenever a better solution is found, the count of the $r$ parameter is set to 1. Finally, the algorithm returns the overall best solution found (step 16 in Algorithm 1).

Notice that, since we are proposing the use of two local search procedures, we find two BVNS variants (BVNS1 based on $LS_1$ and BVNS2 based on $LS_2$).

---

**Algorithm 1:** Basic Variable Neighborhood Search.

---

1 BVNS$(s, k_{max}, r_{max})$
2 **repeat**
3  $k \leftarrow 1$
4  **repeat**
5   $s' \leftarrow$ Shake$(s, k)$
6   $s'' \leftarrow$ LocalSearch$(s')$
7   **if** $f(s'') < f(s)$ **then**
8    $s \leftarrow s''$
9    $k \leftarrow 1$
10    $r \leftarrow 1$
11   **else**
12    $k \leftarrow k + 1$
13  **until** $k = k_{max}$
14  $r \leftarrow$ r+1
15 **until** $r > r_{max}$
16 **return** (s)

---

## 5    Computational Results

In this section, we present the results of several experiments we have performed in order to test the performance of the proposed approaches. All the experiments were run on an Intel Xeon Gold CPU with 80 cores at 2.00 GHz, and with 64 Gb RAM. The operating system used was Debian 9.6. Despite the parallel capacity of the computer, each instance only used one core, hence the execution can be considered sequential.

In order to compare the different proposed algorithmic solutions, we first use a dataset containing synthetic problem instances, and then the best performing solution is tested on a larger problem instance derived from the industry.

The synthetic dataset is composed of 74 instances containing 5 to 25 orders. The number of different items per instance is either 10 or 15, the number of available slots in a workstation is either 2 or 3, and the number of workstations was set to 2, since this is a common configuration in practice. This dataset has been divided into three categories (small, medium, and large) based on different combinations of these parameters.

The problem instance derived from the industry is much larger then the synthetic instances, containing 291 orders.

### 5.1   Results on Synthetic Data

We begin by presenting the results obtained by the mathematical model proposed from Sect. 3 over the synthetic instances using the CPLEX 12.8 solver [1] with default parameters. In addition, we have limited the execution time for a single instance to one hour. The results of this experiment are presented in Table 3, where we report the number of optimal solutions found (#Opt.), the average gap provided by the solver (Gap (%)), and the average CPU time spent (CPUt (s)). The gap is calculated as follows: Gap $= \frac{UB - LB}{UB} \cdot 100$, where UB stands for Upper Bound, and LB stands for Lower Bound.

We can remark that the mathematical model was able to find the optimal solution in 6 out of 20 small-sized instances, while it was unable to find any optimal solution for the medium-sized and large-sized instances. As far as the gap is concerned, we also observe that it grows exponentially with respect to the size of the instances.

**Table 3.** Mathematical model results on synthetic instances

|  | ILP | | |
|---|---|---|---|
|  | #Opt. | Gap (%) | CPUt (s) |
| Small (20) | 6 | 32.68 | 2652.82 |
| Medium (18) | 0 | 84.28 | 3600.45 |
| Large (36) | 0 | 96.99 | 3600.39 |
| All (74) | 6 | 76.52 | 3344.31 |

As expected, the exact mathematical programming approach was able to find the optimal solution for some small instances, however failed to do so for the medium-sized and large-sized instances.

Next, we present in Table 4 the results obtained using the heuristic procedures proposed in Sect. 4.1 and Sect. 4.2 over the synthetic instances.

The constructive heuristic was run 100 times for each instance, reporting the best found solution. We also considered the two BVNS variants described in

Sect. 4.2. As far as the parameters of the BVNS algorithms are concerned, they are set as follows: $k_{max} = \lfloor * \rfloor 2ln(N) + 1$ and $r_{max} = 20$ where $N$ is the total number of orders to be processed and $ln(.)$ is the Napierian logarithm function.

Again, the results are divided based on the size of the instances. For each group we report the number of optimal (#Opt.) solutions (when available from the exact approach), the number of best solutions (#Best) with respect to the problem instance, the deviation with respect to the best solution (Dev. (%)) and the CPU time (CPUt (s)) in seconds.

We can observe that BVNS2 managed to find the 6 optimal solutions retrieved by the mathematical model. In addition, it found the best solution compared to the other approaches over all the small-sized instances, achieving therefore a deviation of 0.00%. Additionally, we can see that BVNS2 found the largest number of best solution over all the instances with a better deviation compared to the other approaches.

**Table 4.** Heuristics results on synthetic instances.

|  | Algorithm | #Opt. | #Best | Dev. (%) | CPUt (s) |
|---|---|---|---|---|---|
| Small (20) | Constructive (x100) | 5 | 16 | 1.37 | 0.00 |
|  | BVNS1 | 6 | 19 | 0.38 | 0.01 |
|  | BVNS2 | 6 | 20 | 0.00 | 0.01 |
| Medium (18) | Constructive (x100) | 0 | 4 | 5.90 | 0.00 |
|  | BVNS1 | 0 | 13 | 2.14 | 0.08 |
|  | BVNS2 | 0 | 14 | 1.54 | 0.08 |
| Large (36) | Constructive (x100) | 0 | 0 | 9.59 | 0.01 |
|  | BVNS1 | 0 | 21 | 2.59 | 0.46 |
|  | BVNS2 | 0 | 24 | 1.78 | 0.46 |
| All (74) | H1 (x100) | 5 | 20 | 6.47 | 0.00 |
|  | BVNS1 | 6 | 53 | 1.89 | 0.24 |
|  | BVNS2 | 6 | 58 | 1.24 | 0.25 |

## 5.2   Results on Realistic Data

In order to estimate the performance of the proposed algorithms for the O3PW in practice, as well as tackle issues linked to the sizing of the picking infrastructure in a warehouse, we consider a very large instance derived from the industry which is composed by 291 orders. While in the previous experiments all the instances were configured with only two workstations, we consider here multiple scenarios where more than two workstations are used in parallel.

Figure 3 illustrates the quality of the solutions provided by the BVNS2, as the largest number of movements of any of the parallel workstations, when considering different numbers of these workstations.

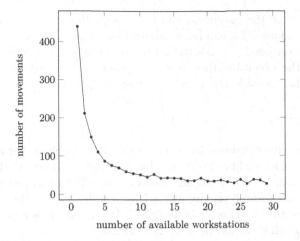

**Fig. 3.** Decrease in largest number of movements when adding parallel workstations for a real problem instance

As expected, the larger the number of available workstations, the shorter the number of maximum movements per workstation, as the workload gets distributed among them. We observe that the largest differences occur when fewer workstation are present, since for example, going from one workstation to two would in the best case divide the workload in two. As more workstation are added, the improvement become marginal. It should also be noted that the results are not optimal, as the mathematical programming approach would not have been able to solve such a large instance.

Next, we highlight the required execution time in Fig. 4.

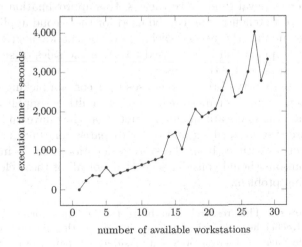

**Fig. 4.** Execution time for the BVNS2 algorithm on the real problem instance

We notice that the execution time increases linearly with the number of parallel workstations. This can be attributed to the local search operator, as the size of a neighbourhood is proportional to the number of parallel workstations. Nevertheless, the execution time remains reasonable in order to be used as a decision aid when considering a warehouse sizing problem.

## 6   Conclusion

In this paper, we have identified a real optimization problem that appears in the context of the order picking process. The problem is named Order Processing in Parallel Picking Workstations and looks for an efficient way of processing the orders arriving to a warehouse by minimizing the largest number of movements made by a workstation. This problem is $\mathcal{NP}$-Hard since it is a generalization of the OPPW [10] which is known to be $\mathcal{NP}$-Hard.

We have proposed a mathematical formulation for the O3PW that can be helpful when we are dealing with small-sized instances and also, we proposed several heuristics based on the VNS methodology that are useful when tackling larger instances. Particularly, we proposed a constructive algorithm, and two Basic Variable Neighborhood Search procedures. All the proposals were empirically tested and compared among them. Finally, we studied the impact of increasing the number of available workstations using our best performing proposed algorithm on a real problem instance.

The practical application of our proposed algorithms is two-fold. The first one consists in providing a day-to-day planning of the operations in the warehouse that can be implemented in order to reduce the time needed to handle a given amount of orders. We estimate this time through the number of movements of containers from the storage area to the picking area, which is a good approximation when the picking and dispatch times are negligible with respect to the container retrieval time. Nevertheless, this approximation may be useful for other types of warehouses and can be used for the second application, which consists of determining the needed picking equipment for a given warehouse. Indeed, this last application can be used to drive the sales of such equipment and it is a critical factor for the industry.

Due to the complexity of real warehouses that contain picking systems, such as the ones described in this paper, we have made multiple simplifications such as considering that containers are single-product, that they have an infinite capacity and that order setup, picking and order dispatch times are negligible when compared to retrieval times from the storage location, just to mention a few. Future contributions should consider several if not all of these elements within the order picking problem.

**Acknowledgement.** This research has been partially supported by Association Nationale de la Recherche et de la Technologie (France) with PhD grant ref. 2017/1525; by Ministerio de Ciencia, Innovación y Universidades (Spain), grant ref. PGC2018-095322-B-C22; and by Comunidad de Madrid and European Regional Development Fund, grant ref. P2018/TCS-4566.

# References

1. IBM ILOG CPLEX 12.8 Optimizer (2018). https://www.ibm.com/analytics/cplex-optimizer
2. Albareda-Sambola, M., Alonso-Ayuso, A., Molina, E., De Blas, C.S.: Variable neighborhood search for order batching in a warehouse. Asia-Pac. J. Oper. Res. **26**(5), 655–683 (2009)
3. Boysen, N., Briskorn, D., Emde, S.: Parts-to-picker based order processing in a rack-moving mobile robots environment. Eur. J. Oper. Res. **262**(2), 550–562 (2017)
4. De Koster, R., Le-Duc, T., Roodbergen, K.J.: Design and control of warehouse order picking: a literature review. Eur. J. Oper. Res. **182**(2), 481–501 (2007)
5. De Koster, R., Roodbergen, K.J., Van Voorden, R.: Reduction of walking time in the distribution center of de bijenkorf. In: Speranza, M.G., Stähly, P. (eds.) New Trends in Distribution Logistics. Lecture Notes in Economics and Mathematical Systems, vol. 480, pp. 215–234. Springer, Heidelberg (1999). https://doi.org/10.1007/978-3-642-58568-5_11
6. Drury, J.: Towards more efficient order picking. IMM Monogr. **1** (1988)
7. Duarte, A., Pantrigo, J.J., Pardo, E.G., Mladenović, N.: Multi-objective variable neighborhood search: an application to combinatorial optimization problems. J. Global Optim. **63**(3), 515–536 (2014). https://doi.org/10.1007/s10898-014-0213-z
8. Duarte, A., Pantrigo, J., Pardo, E., Sánchez-Oro, J.: Parallel variable neighbourhood search strategies for the cutwidth minimization problem. IMA J. Manage. Math. **27**(1), 55–73 (2016)
9. Eisenstein, D.D.: Analysis and optimal design of discrete order picking technologies along a line. Naval Res. Logist. (NRL) **55**(4), 350–362 (2008)
10. Füßler, D., Boysen, N.: High-performance order processing in picking workstations. EURO J. Transp. Logist. **8**(1), 65–90 (2017). https://doi.org/10.1007/s13676-017-0113-8
11. Ganeshan, R., Harrison, T.P.: An introduction to supply chain management, pp. 1–7. Department of Management Science and Information Systems, Penn State University (1995)
12. Hansen, P., Mladenović, N., Todosijević, R., Hanafi, S.: Variable neighborhood search: basics and variants. EURO J. Comput. Optim. **5**(3), 423–454 (2016). https://doi.org/10.1007/s13675-016-0075-x
13. Hansen, P., Mladenović, N.: Variable neighborhood search: principles and applications. Eur. J. Oper. Res. **130**(3), 449–467 (2001)
14. Harland, C.M.: Supply chain management, purchasing and supply management, logistics, vertical integration, materials management and supply chain dynamics. Blackwell Encyclopedic Dictionary of Operations Management, vol. 15. Blackwell, UK (1996)
15. Henn, S., Koch, S., Doerner, K.F., Strauss, C., Wäscher, G.: Metaheuristics for the order batching problem in manual order picking systems. BuR - Bus. Res. **3**(1), 82–105 (2010). https://doi.org/10.1007/BF03342717
16. Henn, S., Wäscher, G.: Tabu search heuristics for the order batching problem in manual order picking systems. Eur. J. Oper. Res. **222**(3), 484–494 (2012)
17. Menéndez, B., Pardo, E.G., Sánchez-Oro, J., Duarte, A.: Parallel variable neighborhood search for the min-max order batching problem. Int. Trans. Oper. Res. **24**(3), 635–662 (2017)
18. Menéndez, B., Pardo, E.G., Alonso-Ayuso, A., Molina, E., Duarte, A.: Variable neighborhood search strategies for the order batching problem. Comput. Oper. Res. **78**, 500–512 (2017)

19. Menéndez, B., Pardo, E.G., Duarte, A., Alonso-Ayuso, A., Molina, E.: General variable neighborhood search applied to the picking process in a warehouse. Electron. Notes Discrete Math. **47**, 77–84 (2015)
20. Mladenović, N., Hansen, P.: Variable neighborhood search. Comput. Oper. Res. **24**(11), 1097–1100 (1997)
21. Nicolas, L., Yannick, F., Ramzi, H.: Order batching in an automated warehouse with several vertical lift modules: optimization and experiments with real data. Eur. J. Oper. Res. **267**(3), 958–976 (2018)
22. Ouzidan, A., Sevaux, M., Olteanu, A.L., Pardo, E.G., Duarte, A.: On solving the order processing in picking workstations. Optim. Lett. (2020). https://doi.org/10.1007/s11590-020-01640-w
23. Pardo, E., Mladenović, N., Pantrigo, J.J., Duarte, A.: Variable formulation search for the cutwidth minimization problem. Appl. Soft Comput. **13**(5), 2242–2252 (2013)
24. Roodbergen, K.J., Vis, I.F.: A survey of literature on automated storage and retrieval systems. Eur. J. Oper. Res. **194**(2), 343–362 (2009)
25. Scholz, A., Henn, S., Stuhlmann, M., Wäscher, G.: A new mathematical programming formulation for the single-picker routing problem. Eur. J. Oper. Res. **253**(1), 68–84 (2016)
26. Tompkins, J.A., White, J., Bozer, Y., Tanchoco, J., Trevino, J.: Facilities Planning. Willey, New York (1996)

# Author Index

Printed in the United States
By Bookmasters